Technically Together

Technically Together

Reconstructing Community in a Networked World

Taylor Dotson

The MIT Press
Cambridge, Massachusetts
London, England

This book was set in ITC Stone Sans Std and ITC Stone Serif Std by Toppan Best-set Premedia Limited.

Library of Congress Cataloging-in-Publication Data

Names: Dotson, Taylor, author.
Title: Technically together : reconstructing community in a networked world / Taylor Dotson.
Description: Cambridge, MA : MIT Press, [2017] | Includes bibliographical references and index.
Identifiers: LCCN 2016052816 | ISBN 9780262036382 (hardcover : alk. paper) ISBN 9780262551229 (paperback)
Subjects: LCSH: Information society. | Communities. | Online social networks.
Classification: LCC HM851 .D66 2017 | DDC 303.48/33--dc23 LC record available at https://lccn.loc.gov/2016052816

Contents

Acknowledgments

A book is never a solitary accomplishment. Even though a single name appears in the byline, this work is the culmination of countless hours of conversation, debate, and much needed encouragement provided to me by dozens of people. That being said, responsibility for this book's various flaws and oversights rests solely with me. All the support and helpful comments I received were ultimately interpreted and applied through the filter of my own biases and cognitive limitations.

My mentor and friend Ned Woodhouse has been instrumental to this project's realization. Without his care and dedication when reviewing drafts and willingness to listen to and help improve half-baked and naïve ideas this book would have been a much more flawed work. I do not know if I turned out to be even half the writer or scholar he hoped I would become when he took me on as his student, but I am grateful to have been under his tutelage. His guidance has left an indelible mark on my thinking concerning matters of technology and society.

I would like to also acknowledge Langdon Winner, Abby Kinchy, Mark Mistur, David Brain, and three anonymous reviewers for their helpful comments and incisive criticisms. I am thankful that they took me to task for occasionally shallow analysis, ignorance of important literature, and other lapses. I needed it. I am especially appreciative of Langdon Winner's constant encouragement and belief in me as a scholar. It has meant a lot to me. I am likewise grateful for my interactions with Linnda Caporael over the years. Talking with her has always been a pleasure and mind expanding. Her influence on this book, though more informal, has been significant. I also want to thank Katie Helke and MIT Press for their belief in the merits of *Technically Together* and all their hard work in getting it through the publication process.

I am indebted as well to several individuals who studied and researched alongside me in the Department of Science and Technology Studies at the

Rensselaer Polytechnic Institute. I am thankful for their assistance in my attempts to maintain some semblance of sanity and work-life balance. I greatly appreciated the social support offered to me by Nate Fisk, Dan Lyles, Karin Patzke, and Michael Bouchey; for the friendship and academic collaboration provided by James Wilcox, Michael Lachney, and Ben Brucato; for Colin Garvey's meditation sessions, willingness to chat with me about almost anything, and wonderful sense of humor; and for the caring camaraderie over the years of Kate Tyrol, Wynne Hedlesky, Khadija Mitu, Lindsay Poirier, Laura Rabinow, Ellen Foster, Joe Datko, Jess Lyons, Sabrina Weiss, Ross Mitchell, Gareth Edel, Jon Cluck, Guy Schaffer, Erik Bigras, David Banks, Kirk Jalbert, Brian Callahan, Sonia Saheb, and Ali Kenner.

I am grateful for my friends, without whom I would not have had the psychic reserves to finish this book. In no particular order, I thank David Morar, Franz Mathes, Alex Sobolev, Antonio Chavarria, Guilhem Werbelow, Sally and Mike Catlett, Carla Spina and Chris Gillespie, Kristian Isringhaus, Stephen Bradley, Liz Westcott as well as Greg and Alyla Goldman. I also would like to express my appreciation to the good people at both the First United Presbyterian Church in Troy as well as the Albany Curling Club. They provided me with not merely a much needed boost in social belonging but important insights into the practice of community as well.

Even though I have largely fallen out of touch with my former roommate, Matt Finnell, he deserves my gratitude as well. I began to realize and develop my interest in the social world, particularly community, partly as a result of conversations with him in the student union building of New Mexico Tech when we were both undergraduates.

Completing *Technically Together* would not have been possible without my family. The support of my parents, Richard and Renee, and my in-laws, Tabea and Wayne, was a big help on the days that I felt down. I am grateful for the innumerable small ways that they reminded me that my book's topic and argument mattered, not only to them but also to the rest of humanity. My brothers, Tom and Stu, have been equally important in this role. I thank them for being there and listening to my complaints, hopes, and worries. I am likewise grateful for the encouragement provided by my sister-in-law, Cris Thomas, and the delightful distraction offered by my nephew, Antony, when I stayed with them for several weeks of intensive writing during the summer of 2014.

Finally, I could not have done any of this without the loving companionship of my wife, Rachel. Her sunny disposition, unwavering belief in me, and firm insistence that I stop working so hard and take a break once in a while made this book possible. I am truly blessed to have a spouse like her.

1 The Question Concerning Technology and Community

Too seldom is it considered that the character of social life could be different than it is today. Far too many people act as if the social status quo were as natural as the daily rise of the sun, as if changes to the practice of community were inexorably imposed on humanity by technology rather than enacted by people through technologies. Network society living is too often treated as either a blessing to celebrate or a curse to bear rather than something to be reconstructed. While the occasional magazine article sounds the alarm about Facebook potentially "making us lonely,"[1] popular discourse seems permeated by professional social analysts assuring citizens that technological changes to social life are nothing to worry about. Sociologist Keith Hampton, for instance, has maintained that the "constant feed of status updates and digital photos from our online social circles is the modern front porch," insisting that "those we depend on are more accessible today than at any point since we lived in small, village-like settlements."[2] According to Internet scholar Yochai Benkler, "Social norms and software coevolving to offer new, more stable, and richer contexts for forging new relationships beyond those that in the past have been the focus of our social lives."[3]

We? A more stable and richer context for *our* social lives? To anyone who has ever felt lonelier after checking Facebook than before, such claims read like bromides from Dr. Pangloss. They seem to assume away significant problems with contemporary social belonging, implying that "we" are all on our way to living in the best of all possible worlds. Anyone who has enjoyed a summer day on the front porch with friends and neighbors, however, would likely not equate it with social media. Try getting a Facebook status update to help move a couch or stay for dinner. It is doubtful that people who are dissatisfied with the current level of community in their lives take such proclamations seriously. To them, Benkler's and Hampton's paeans to digitally enabled social networks read like reverse adaptation:

they appear to call for citizens to adapt their expectations for social life to what dominant technologies can offer rather than demand that they be refashioned to better enable other ways of being and relating.[4] The first task of this book is to challenge the common refrain that people ought to stop worrying and learn to love social networks: What is lost and who loses in the shift toward network-based social life?

If social media and personal communication devices have actually delivered on the social wonders promised by their champions, one would not know it from the available data. One study created a stir when its authors calculated that the number of people having no close confidants had tripled over a few decades.[5] Critics have since argued that the study's results were erroneous, revising the estimate to a more modest fifty percent increase.[6] Accurately measuring such changes, in any case, is hindered by the fact that people's interpretation of what it means to have a confidant shifts over time and in response to technological changes. On the other hand, a 400 percent increase between 1988 and 2008 in citizens taking antidepressants seems to suggest that at least some people may not be receiving the kind of belonging, intimacy, or psychosocial support they need.[7] Indeed, a recent survey found that over a third of American adults over forty-five meet the criteria for "lonely" on UCLA's loneliness scale.[8]

Although it is unclear to what extent people may be lonelier today than in the past, most scholars recognize that social activity has become increasingly individualized and fragmented; that is, people experience their sociality more and more often through diffuse, atomistic networks of specialized social ties rather than in bounded, densely woven, place-rooted, and economically, politically, and morally rich communities. They experience social connection as individuals accessing and moving through technological networks rather than as an almost indelible feature of everyday life. Individuals more often correspond via Facebook or attend ad-hoc "meetups" than join local associations or frequent neighborhood cafés. Social ties are more diffuse and segregated; community appears more as a friends list than as a place to which one can point. *Technically Together* offers a critical inquiry into this shift: How do technological arrangements make certain facets of traditional community life more difficult to realize? Which technologies help reinforce the social status quo? What technical, political, economic, and cultural changes would enable citizens to weave their fragmented, diffuse social networks back together into place-rooted communities?

The Community Question

The slide toward more individualistic forms of social connection has been characterized by a number of sociologists. Robert Bellah and his coauthors, for example, found that Americans have become increasingly incapable of coherently describing or understanding their social obligations.[9] An ever more dominant language of individualism leaves people unable to comprehend their connections outside the logic of utilitarian self-interest or expressive self-realization. These scholars worried that such linguistic changes would weaken the practices of commitment that sustain marriages, families, and communities, because such practices can seem irrational when analyzed starkly in terms of short-term personal costs and benefits. Group belonging, moreover, has been increasingly sought through *lifestyle enclaves*, groups organized around shared personal interest rather than in terms of broader collective goals or responsibilities.

Robert Putnam's impressively thorough *Bowling Alone* painted a similar picture of twentieth-century communality.[10] Although sometimes mischaracterized as claiming that Americans had become totally socially isolated, Putnam actually analyzed the decline of certain forms of sociality, not the quantity of social ties. Americans are not really "bowling alone" so much as decreasingly doing so in local bowling leagues. They instead bowl sporadically with different sets of friends. Moreover, they are less likely to spend time with neighbors, entertain at home, join clubs or unions, or participate in local politics. Putnam's work systematically charted the increased social fragmentation of multiple spheres of daily life and pointed to suburbanization, broader cultural changes, and television as contributing factors.

General levels of social trust, neighboring, and membership in formal organizations have continued to decline since Putnam's book was published in 2000; while volunteering and joining increased among young Americans somewhat after 9/11, this uptick in civic activity has occurred primarily among upper-middle-class whites.[11] At any rate, the slide away from the regular, socially leveling interactions characterizing local associational life and toward less coherent and more fragmented social networks signals the demise of an important form of communality. Journalist Marc Dunkelman describes this as a loss of the "middle ring" of social life. While people remain enmeshed in an inner ring of intimate bonds and connected to an outer ring of online and distant ties, neighborhood- and township-level relationships are in decline.[12]

The social network paradigm of community studies, largely developed by sociologist Barry Wellman, has confirmed this shift in social structure *toward networked individualism.* From his 1970s studies of Toronto social networks to more recent collaborations, Wellman has depicted social ties as increasingly geographically dispersed, more loosely knit, more specialized, and increasingly based on shared personal interests.[13] Social belonging less frequently occurs through local, dense, and interweaving webs of bonds. Networked individualist connections are created and maintained only insofar as they support the bilateral exchange of some element of social support rather than because of local solidarity or the need to sustain a club, an association, or a neighborhood. A networked individualist maintains certain connections for a shoulder to cry on, others for going out on the town, and still others for professional advancement, but rarely do these connections intersect or even know of one another.

Social network theorists typically part ways with scholars like Bellah and Putnam, insisting that there is little reason to worry about networked individualization. They depict the increased bypassing of intermediate ties and scales of organization via networks as an evolution of community—neither good nor bad—rather than a decline. A case in point is Claude Fischer, who has maintained that not much has changed in the last forty years concerning American's personal connections.[14] Using national survey data, he demonstrated that neighboring and entertaining at home has been replaced with going out with friends or chatting online. Family time likewise happens less around the dinner table; members increasingly interact with one another while driving to appointments or activities. As insightful as Fischer's research is, it does not really answer the concerns posed by Putnam and Bellah. Its narrow focus gives the impression that qualitative changes in how people spend time with each other can be ignored as long as the quantity of social contact stays roughly the same.

Other scholars have been even more up front with their optimism, appearing to depict networking as a nearly unalloyed improvement compared with older forms of social life. Lee Rainie and Barry Wellman, for instance, have extoled networked individualism's virtues as a revolutionary "social operating system." They argued that networked community "offers more freedom to individuals than people experienced in the past because now they have more room to maneuver and more capacity to act on their own."[15] Similarly, sociologist Eric Klinenberg has described living alone and without strong social obligations as a modern virtue, for it "liberates us from the constraints of a domestic partner's needs and demands, and permits us to focus on ourselves ... [and] helps us discover who we are, as well as what gives us meaning and purpose."[16] This response to a decline in the community life described by Putnam and Bellah amounts to saying "good riddance"—a sentiment increasingly embraced, by Americans in particular, over the past decades.

Beyond Networked Individualism

Why are the consequences of qualitative changes in the practice of togetherness so quickly downplayed when psychological data indicate that Americans are becoming more narcissistic, self-centered in their moral reasoning, dismissive of intimate relationships, and less empathic?[17] Why is it that many social network scholars appear to presume that they know what is socially liberating for the rest of technological civilization? Ideas like liberty, meaning, and purpose are matters of value, and hence politics, rather than ostensibly objective social facts. What *freedom* means, to take one example, has been debated by political theorists for millennia and is inexorably conditioned and constrained by the culture one grows up in. It is not something whose true nature can be discovered through a social survey or ethnographic study. Other highly value-laden terms, including community, are no different. Although social network analysis can help elucidate the various ways through which people might seek belonging, it cannot determine what belonging *ought to mean* to them. Concerning the latter question, social network scholars have no special epistemological authority. In making grand universalizing statements about what community is and what gives contemporary persons meaning, one risks assuming away these issues' political and moral complexities. Such questions inevitably evoke disagreement about the character of the good life and how it should be provided.

The aim of this book is to recast the question concerning community and technology as a political one. Rather than believing that community is something discovered, I frame it as something to be debated and shaped through political action: Who benefits from which technologically enabled forms of togetherness? Which ways of providing social belonging are desirable? Which technologies enable and which constrain various types of interpersonal life? The widespread neglect of such questions virtually guarantees that citizens remain passive observers to large-scale alterations to social life rather than agents of change.

My analysis begins with a rejection of technological fatalism. Carolyn Marvin once pointed out that people too frequently presume that "new technologies will make the world more nearly what it was meant to be all along."[18] Technologically driven changes to community life are too often portrayed as inevitable stops on the grand railroad of progress toward modernity, for better or worse. Rather than simply lament or celebrate where contemporary social life *seems to be headed*, I focus on *how it could be otherwise*. How could technological societies better enable thicker forms of togetherness more of the time?

While there is nothing preordained about the character of "modern" living, most people probably feel as individually helpless in changing the social status quo as they do in attempting to eliminate their carbon footprint. Practices such as networked individualism appear to be virtually locked in to the fabric of everyday life.[19] Yet, technological changes only *seem* unalterable because they are supported by a whole host of supportive policies, economic arrangements, and entrenched patterns of thought. What sociotechnical factors render alternative modes of belonging more difficult to enact? At the same time, major social and technical upheavals have almost never been the product of a conscious and foresighted steering of technology. Societies typically sleepwalk through the process of sociotechnical evolution.[20] Rarely, if ever, has the question "What might be the effects on social life?" guided decision making regarding which technologies ought to get funded, designed, and deployed. How could citizens more effectively govern technological change with respect to its potential effects on togetherness?

Why the Individuation of Social Life Matters

The fact that networked individualism is becoming ever more entrenched into the sociotechnical context of everyday life without conscious, collective deliberation is important not just because it is a clear violation of democratic principles but because it affects people's welfare. What I call *thick community* is associated with several aspects of physiological, psychological, and cultural well-being.[21] People who are well integrated into social communities or localities high in social capital tend to live longer, suffer fewer psychological problems, and enjoy greater political democracy, among other benefits. This may be because communal integration helps prevent a sense of loneliness, alienation, or purposelessness. In any case, the feeling that contemporary societies are unwell probably drives the ongoing perception that community is in decline, despite the fact that it lives on, in some sense of the word, via fragmented social networks.

At the same time, many environmental scholars have suggested that growing resource scarcity and global ecological disruption are crises that technological societies are unlikely to innovate their way out of.[22] Rather, adequately adapting to or steering away from the brink of global economic and ecological catastrophe will require making significant sociocultural changes, including shifting toward more communitarian economies and social formations.

Moreover, political thinkers have argued that realizing freer and more just societies lies in sustaining diverse and strongly democratic communities.[23] Sociologist Robert Nisbet long ago found that the emergence of European totalitarianism paralleled the dismantling of social organizations lying at scales between that of the state and that of individuals, weakening what had provided a countervailing force to state power.[24] In fact, Nazi leaders consolidated and assured their control partly through outlawing independent labor unions, mutual aid organizations, and other civil associations that might have offered fertile ground for the seeds of resistance to an increasingly intrusive centralized state. The threat to free and open societies today, of course, is no longer outright totalitarianism, at least outside a few embattled countries. Nevertheless, the less brutal but still significant public harms produced by unchecked corporate power persist partly because of the relative weakness of civil society as a countervailing political force. The pervasive cultural overvaluation of individual liberty (especially economic) at the expense of democracy, coupled with the individuation of social life, leaves citizens poorly equipped to defeat better organized and funded elites in political conflict. The fact that most people's idea of politics is limited to largely individual and insubstantial efforts such as occasional visits to the election booth, slightly more conscious and circumspective consumer purchases, and/or forms of online "clicktivism" both signals and reinforces the decline of the democratic freedoms achievable through a communitarian civil society.[25]

Regardless, one need not go so far to appreciate the value of thicker forms of social connection. Many of those reading this book might already wish that their environment were more conducive to getting to know neighbors, facilitating leisurely socializing with friends, and engendering a sense of belonging to locality. It is not uncommon to hear citizens lament the extent to which their social lives are driven by sporadic meetups with fragmented friend groups and countless hours surfing online social networks. For some, these are imperfect and short-lasting substitutes for thicker forms of communal belonging (see box 1.1).

Whither Technology Studies?

Despite the importance of social belonging to people's welfare, topics such as community have rarely been the target of science and technology

Box 1.1

Networked Individuals: Thriving or Coping?

In *Networked*, Lee Rainie and Barry Wellman presented Linda Evans as the embodiment of learning to thrive in a networked world. Depressed and only recently back at work after her divorce, Linda successfully operated a range of online and offline social networks to put her life back together. She joined a support group at her church, eventually developing a deep online relationship with another recent divorcé named John. She advanced herself professionally and financially through online degree programs and by accessing Internet resources on investing. Moreover, Linda started running an online support group for people with myasthenia gravis, finding meaning in helping those anxious about their future or rendered housebound by illness. Rainie and Wellman depicted people like Linda as having the right "combination of talent, energy, altruism, social acuity, and tech-savviness" necessary to pull themselves up by their bootstraps and succeed socially, personally, and professionally as networked individuals.[26]

The BBC documentary *The Age of Loneliness* complicates and challenges this view of social belonging in a networked age. Interviewees included a college student named Isabel, a more than one hundred-year old woman named Olive, and people from nearly every life stage in between. They did not see themselves as "thriving" in an age of networked living. On the contrary, most described themselves as merely coping. Interviewees made ample use of social networks: volunteering, organizing weekly mom's meetings through Facebook, Internet dating, and attending computer classes, but doing so yielded mixed results, and most remained lonely. For Isabel and a divorcée named Kylie, online networking exacerbated feelings of loneliness. They perceived status updates and profiles as other people's "highlights reels," carefully curated depictions of fun and excitement that make others feel more alone in comparison. To Jane, a single woman in her forties, online dating sites mainly provided new opportunities for rejection. Others found it difficult to make and sustain new friendships, expressing the wish that someone would just stop by regularly to chat; however, one "can't make people come and ring the [door]bell." Emily, a stay-at-home mother who relied on supermarket cashiers for her daily social contact, lamented lacking a local extended family network like the one that had supported her own mother.[27]

The experiences depicted in *The Age of Loneliness* doubtlessly result from a combination of the interviewees' circumstances and their personal characteristics, such as mental illness. Nevertheless, in light of their struggles, the growing expectation that individuals be solely responsible for cultivating their own social communities starts to seem unfair, if not uncaring. Descriptions of winners such as Linda Evans, on the other hand, begin to read like digital age "rags to riches" stories. Similar to the tales told about immigrants who arrived in America with only a dollar in their pocket and became wealthy by the sweat of their brow, winners' stories imply that the resources for success are simply there for the taking. Failure to amass either wealth or social ties becomes framed as a personal, not a societal, failing.

studies (STS) research. The examination of how the technical is sociopolitically shaped tends to overshadow questions regarding the consequences
of sociotechnical change for the average Joe or Jane. Although Anique
Hommels has well illustrated how sociopolitical forces render some urban
infrastructures obdurate or difficult to change, her analysis devoted little
attention to the forms of life that such infrastructural obduracies promote
or inhibit.[28] Sociologist Thomas Gieryn echoed what urban theorists from
Jane Jacobs to Kevin Lynch have argued for decades when observing that
"buildings stabilize social life"[29] but did not ask which buildings were up
to the task of stabilizing what kinds of social lives. Additionally, "sexy" or
esoteric technologies, like synthetic biology or nineteenth-century bicycles
often receive equal—if not more—attention as the more mundane technologies shaping everyday life. One of the goals of this book is to demonstrate
what technology scholarship aimed at addressing the problems of ordinary
people could look like.

My intention is to provide sociopolitical analysis for the benefit of
those who seek to reconstruct contemporary societies to better provide
more multifaceted and vibrant forms of belonging. As such, my goal is not
really to convince staunch advocates of networked individualism that they
are misguided, though I hope the following analysis helps some of them
recognize how community life is hollowed out if marketlike networking
crowds out other modes of being. Rather, my purpose is to help aspiring
communitarians—as well as those not yet sure of how they feel about networked sociality—better see how technological societies stymie the development of alternative forms of togetherness and how they might begin to
strategize toward a different future.[30] I do not simply catalog the problems
technological developments have created for social life but also examine
the possibilities for realizing sociotechnical change.

These features—and others—make *Technically Together* atypical within
the field of STS. My focus on the consequences of technologies for people's subjective experience and well-being differs from standard social con-
structivist approaches, which integrate those experiences mainly insofar as
they explain scientific fact production or technological innovation. Paying
attention to the built environment, despite its importance for everyday life,
is also unusual for STS. I could count no more than a handful of papers
focusing on urban technologies in the field's main journals in the last five
years, though talks regarding urban life at the STS annual conference have
become more common since 2014. Urban technologies have historically
been left to planning and architectural scholars to analyze. Likewise STS
scholarship rarely attempts to imagine or strategize toward alternative

futures. Indeed, emphasis is usually placed on constructing histories of change and on mapping the politics of contemporary controversies. Normative recommendations are usually made in the final chapter, if at all. The tendency to shy away from prescribing what could be done and to focus on describing what has already happened means technology scholarship provides more academic benefits vis-à-vis theory building than practical gains. STS research is too often of limited helpfulness to anyone not holding a university faculty position.

Preview

In the next chapter, I continue my analysis of the networked sociality paradigm for understanding the interrelation between community and technology. What are the politics of networked individualism as both a *phenomenon* and a *theory*? I show that the practice of networked individuals entails both winners and losers, especially in how belonging shifts from being a public good to a private responsibility. Such politics, however, have generally received little academic attention. I partly attribute this lack of attention to how technological change is conceptualized in texts by network scholars. Although they recognize that the validity of networked individualism *as a theory of community* is ensured by technological changes that help establish it *as a dominant social phenomenon*, they tend to underemphasize how such changes are sociopolitically accomplished. As a result, Wellman and other network sociologists can come across as justifying or normalizing technological changes to sociality as natural or inevitable, limiting the scope of debate to exclude the imagining of alternatives to the status quo. Such a portrayal is not simply the result of dispassionate social scientific analysis but rather amounts to an implicit form of political advocacy.

In chapter 3, I develop an improved framework for understanding community. Too often, the concept is reduced to a single variable, such as social networking or the sense of belonging. Yet, approached differently, the corpus of research on social connection appears as a rich trove revealing community's multiple facets and manifestations: symbolic and rhetorical markers of collectiveness, civic self-governance, and the skills and dispositions needed for longer-term social commitments, among others. This approach enables the characterization of different instantiations of social phenomena as *thick* or *thin* with respect to the multiple dimensions of community, allowing for a more nuanced analysis of how communitarian practices are affected by sociotechnical change. I describe how many instances of contemporary communality, including the phenomenon of networked individualism, are *relatively thin* in comparison to what has existed at other times and places and what could exist in future technological societies. Finally, the framework breaks away from stale arguments that reduce the debate to a comparison between idealized and mostly hypothetical visions of the ostensibly intolerant rural hamlet and the purportedly liberating but anonymous city. Social belonging is a more multifaceted and diverse category of phenomena than such dichotomizing views presume. Even though I strongly criticize networked individualism, *my purpose is not to present it as the opposite of community but as a genuine instantiation of it,* albeit one that I find to be relatively thin in regard to several dimensions of communality.

In chapters 4 through 6, I extend to the phenomenon of community political philosopher Langdon Winner's observation that "different ideas of social and political life entail different technologies for their realization."[31] Which technologies help to naturalize networked individualism? What features of different artifacts, techniques, systems, spaces, organizations, and infrastructures make them more amenable to different dimensions of thick community? I subsume a far broader range of objects under the category of *technology* than is typical in common understandings of the word. A narrow focus on devices and tools can blind one to the equally important and supportive role of techniques, urban form, organizations, and large-scale infrastructures. Strategies for sleep-training infants and structures for organizing energy provision or youth sports are as technological as cell phones and neighborhood streets. The following are among the questions raised regarding potentially communitarian technologies: Does it gather social activity in a place or promote diffusion? Does it encourage the negotiation of conflict or its avoidance? Can it help evoke the sense that one exists within a coherent social group? Is it premised on sharing or on individualistic consumption? Does it draw users into webs of material and/or economic interdependence? Through this analysis, I illustrate that the usual suspects (e.g., television and suburbia) are far from the only or even the most significantly anti- or thinly communal technologies.

In chapters 7 through 9, I reconsider the same technologies from a very different angle. What are the barriers to constructing and reconstructing technologies to be more supportive of thick community? In other words, what contributes to the obduracy, or resistance to change, of contemporary social patterns, and how could that obduracy be lessened? Rather than pretend that individuals have the capacity to rearrange their inherited technological contexts to accommodate the type of togetherness they desire,

I outline some of the main social, political, economic, and cultural factors that make such efforts difficult. For instance, if the "cry-it-out" method for sleep training infants is a psychosocial technology that helps ingrain individualistic beliefs and dispositions, then determining what barriers stand in the way of more relationally communal sleep training should matter to would-be communitarians. The feasibility of alternative sleep-training methods depends on the time and scheduling constraints placed on contemporary parents. There are sociotechnical reasons why "crying it out" might appear to be the only practical approach for the average parent. For competing strategies to become viable, workplaces would need to become much more flexible regarding alternative scheduling, and other institutions would need to provide new parents with other kinds of support.

The most significant barrier to thicker communities, however, may be ignorance, for an increasing fraction of people have had little experience with thicker forms of belonging. Their understandings will be limited by generational amnesia as well as by the parallel rise in social anxieties. Indeed, more and more people view social interactions and obligations as a source of trepidation, with many younger people preferring to text rather than phone friends or to order pizza from a website so as to avoid talking with a stranger.[32] Again, many peoples' lessened expectations and habits regarding social life are often the result of their "reverse adaptation" to the life patterns their technological context best supports. I hope this book can provoke both citizens and scholars to reconsider taken-for-granted stories and supposed truisms that currently constrain thinking about what community could mean.

In chapter 10, I examine how technologies that risk harm to thick community could be more adroitly governed. What changes in decision making concerning technological development could permit a more conscious and reflective approach to large-scale sociotechnical changes to social life? This chapter extends research regarding how to productively cope with the uncertainties and complexities present in novel sociotechnical undertakings. Political scientists have found that decision makers are able to "avert catastrophe" insofar as they proceed carefully and in light of experience—that is, through "intelligent trial-and-error."[33] Most studies have applied this framework to technologies that pose significant financial, environmental, and human health risks. How might it be applied to emerging technologies that could negatively affect the practice of community, namely driverless cars and companion robots? At the same time, intelligent trial and error is not only applicable to risky technologies but to large sociotechnical undertakings more generally. How have builders of communitarian technologies used intelligent trial-and-error strategies to more reliably deliver on their promises? I consider a recently developed neighborhood in Freiburg that has achieved high rates of walkability through the involvement of citizen-led building collectives in a broader process of "learning while planning." The success of this German neighborhood provides a lesson in how to improve the application of communitarian urban design ideas, including new urbanism, elsewhere. A comparison of the fortunes of two very different New York food cooperatives further illustrates how effective trial-and-error learning can influence the success of communitarian endeavors. At the same time, the barriers to intelligently steering technological development loom large. I explore what could be done to lessen those barriers.

Technically Together offers a very different perspective on networked individualism than is typically presented. I do not simply emphasize the dimensions of community life poorly supported via networks but go on to describe the thinning out of sociality as highly contingent, not preordained. Everyday life is the product of numerous and vast sociotechnical systems: people, institutions, artifacts, built environments, infrastructures, and beliefs.[34] The thinning out of togetherness has happened through the gradual replacement of sociotechnical components that had previously supported thick community with more individuating versions of them. The substitution of vibrant urban areas with suburbia is one example, a process accelerated by the amenability of other sociotechnical systems to suburbanization—including financing, transit networks, planning policy, and cultural ideas about the good life.

Caveats

Readers, however, should not expect me to offer a silver bullet solution for the problem of community decline, for there is no single technology or policy that would bring about revolutionary changes in how societies provide communal belonging. Large-scale sociotechnical changes, despite occasional revolutionary rhetoric, happen ploddingly over the course of years, decades, and generations. Reconstructing societies to be more thickly communitarian would require the incremental and piecemeal replacement of much of the scaffolding of everyday life, as was the case for the networked individualization of social connection. Not only do material components, including neighborhoods and transportation systems, need to be altered but so do less tangible elements, such as zoning codes, systems of governance, practices of child rearing, and cultural beliefs.

Neither do I nostalgically pine for romantic idealizations of community. References to the "good old days" do not appear in the following pages. Even though everyday life has been more communitarian at other times and places, those eras and locales had other problems. As Putnam noted, the "golden age" of civic activity in the United States coincided with high levels of structural racism and sexism.[35] At the same time, thick community life should not be relegated to the dustbin of history simply because some previous societies could have practiced it better. To think otherwise ignores the existing and potential diversity of the phenomenon. Citizens banding together to fight racism in their cities and to encourage their neighbors to be more tolerant are as much manifestations of community as the more undesirable instances that come to mind. In any case, it is better to focus on the future. How could future technological societies be more communitarian *and socially just*?

Furthermore, I do not claim that communities could or should be radically thickened overnight. As with any other complex sociotechnical endeavor, efforts to thicken social life should proceed intelligently, which often means incrementally. Realizing the best features of thick community while avoiding potentially undesirable side effects will take some combination of thoughtful and democratic deliberation, gradual learning from experience, and other elements of intelligent trial-and-error learning. Even though I dedicate considerable attention to the barriers to realizing more thickly communitarian societies, I do not provide a foolproof or exhaustively detailed roadmap. The best strategies for overcoming the relevant barriers will be borne out of experimentation by activists and others. I only aim to help readers begin to think more deeply about how the sociotechnical world presently limits opportunities to realize thick community and to consider what might begin to be done about it.

Finally, although I have made efforts to draw on a diverse range of examples and data, the analysis is presented from a North American's point of view. Its applicability to other nations depends on the extent to which they have moved toward American-style networked individualism. In any case, I hope my depiction of the social consequences of North American technological changes can serve as a warning to people living in other nations. Such readers should pay particular attention to the various political, economic, and cultural factors that I depict as aiding the momentum of the technologies of networked individualism, if they wish to protect themselves against similar changes to social belonging in their own countries.

I contend that people should neither pine for a mythic past nor fatalistically accept the status quo but should pursue collective action to realize more desirable futures. Technological societies that better support more multi-faceted communities for a broader range of their members would be possible if citizens quit sleepwalking through the process of technological change and demanded more from their technologies. By the end of this book, I will have sketched out the beginnings of how such societies could be realized.

2   The Politics of Networked Individualism

Social science is inexorably political. While STS scholars have demonstrated how culture and politics have permeated technoscientific endeavors such as missile guidance systems, failed public transportation systems, and the detection of gravity waves,[1] there is little reason to suspect that the social sciences themselves are immune to the same sociopolitical forces that affect science and technology. Indeed, social scientists would be unique among human beings if their perspectives were not even subtly shaped by their cultural experiences and political commitments. In any case, the incorporation of cultural assumptions and value commitments into the minting of new social facts helps drive the *naturalization* of what are in actuality socially constructed realities.[2] When caught up in this process of naturalization, social science no longer merely describes reality but contributes to its reconstruction—insofar as it leads some citizens to cease to agitate for change and to adapt to the status quo. In characterizing such assumptions and commitments with respect to networked approaches to sociality, I aim to denaturalize and, in turn, repoliticize networked individualism.

My argument is similar to sociologist Leonard Nevarez's analysis of how dominant understandings of "quality of life" influence debates about the greater good in contemporary societies. Nevarez described the popular and academic discourses centered on the concept of "quality of life" as reflecting and reinforcing individualistic conceptions of well-being, hiding social and political divisions and inequities, and limiting the range of thinkable actions to "accommodating and finding satisfaction from the choices and constraints presented by the world before us."[3] As a result, thinking about the greater good becomes limited by exclusively individualistic and consumerist values; collective, socially just, and democratic conceptions of the good are, in turn, buried. Hence, I ask: What are the limitations of the network-based *theory* of community with respect to the politics of networked individualism *as a social phenomenon*? In what ways do assumptions within network scholarship seem to foreclose debate about alternative technological futures? What is the political character of these assumptions? My exploration of these questions will show that discourse around networked individualism often risks limiting the range of thinkable actions regarding technology and community. Simply put, scholarship on networked individualism has politics.

There is nothing wrong with using network metaphors and methodologies to understand social life per se. Indeed, even Putnam's work on social capital makes use of the concept of social networks.[4] My focus is mainly on occasions when discourse regarding networked individualism does political work: naturalizing the fragmentation of community, downplaying the production of winners and losers, or depicting the individualization of sociality as an inevitable feature of modernity. This chapter is not meant to be a diatribe against network sociology writ large but a provocation: a challenge to reexamine the politics of networked individualism.

The Winners and Losers of Networked Individualization

Politics can be defined as the answer to the question "Who gets what, when, and how?"[5] If large-scale technological changes influence "Who gets what kind of community, when, and how?" then such changes amount to a kind of politics regarding social life. They are tantamount to political regulations achieved via technical means. Directing one's attention to the "who," "what," and "how" of the social phenomenon of networked individualism helps clarify its political dimensions, which tend to be underemphasized in many networked individualist theories of community.

Under networked individualism one belongs to multiple fragmented communities, which themselves owe their existence only to the temporary coalescing of otherwise unrelated social ties. Community in a networked world is simply the extent of the personal ties that individuals have managed to amass for themselves. Networked individuals attempt to obtain a sense of social belonging through their Facebook connections and the contact list of friends that they text to meet up for coffee. Networked sociality occurs across space rather than in places. Socializing entails driving or communicating across considerable distances rather than simply showing up at a local pub or walking in one's neighborhood.

For many people, however, "networked community" may be as much a problem as a solution with respect to contemporary belonging (see box 1.1). People differ regarding the extent to which they feel it satisfies their needs for social connection. Why is this the case? One possible reason is that, under networked individualism, belonging is more of an individual responsibility than a public good. Political scientist Robert Lane noted that "What was once *given* by neighborhood and work now must be achieved ... [Contemporary market society] makes friendship hard work."[6] While some people have little trouble creating and sustaining a social network, the constant organizational labor of parties, dinner dates, and meetups requires a level of social aptitude and energy that others can find discouraging. In networked societies, people's locales rarely provide repeated, unplanned, and mostly pleasant interactions with proximate others, which, in turn, form the social soil from which friendships and other communal relationships can blossom.

In the shift from being a public good to a private responsibility, market-like language has increasingly come to characterize discourse concerning social belonging. Network theorists have described networked communities as "portfolios" of social support that are assembled by "entrepreneurial operators."[7] Likewise, community informatics scholar Michael Arnold has argued, regarding local Internets, that "a Community Network should not be theorised as a public good infrastructure supporting ... community. In an important sense a Community Network is *a resource for building private assets*"[8] (emphasis added).

This shift in discourse, of course, partly reflects on-the-ground changes: tech firms such as Facebook and Apple, providers of social gathering spaces (e.g., Starbucks), and gated community builders play an increasing, and increasingly profitable, role in helping entrepreneurs of social connection earn a suitable rate of return on their social portfolios. Belonging depends evermore on citizens' ability to *afford* special gadgets to access social networks or the financial means to live in certain neighborhoods.

The shift from social connection being a public good to a private responsibility affects the character and benefits of belonging. One classic study found that the unconditional forms of mutual support characteristic of thick communities help people weather the stresses of everyday life, contributing to better health. As one small Pennsylvania town became less communal and more individualistic, rates of heart attack and old-age dementia quickly rose in tandem.[9] Similarly, as developmental psychologist Susan Pinker has outlined, studies of the spry health and longevity of residents in centenarian-laden places like Sardinia attribute their high levels of well-being to the strength of local webs of supportive, long-lasting face-to-face ties more than to any other factor.[10] Even though networked individualism offers more choice regarding with whom one socializes when compared to thick community, it may come with risks to citizens' physical well-being.

The privatization of social connection, furthermore, directly discriminates against certain populations. Those lacking the financial, social and cognitive resources *to network*, such as the mentally ill, homeless people, and those challenged in their social skills and charisma, are put at a disadvantage. Such groups were disadvantaged, of course, prior to the dominance of networked individualism, but the privatization of community only increases the social disparity between them and others. This is because the moral logic of networked individualism makes having some kind of "good" to offer—whether it be social support, charisma, or something else—a prerequisite for social connection. On the other hand, thick communities, at their best, unconditionally integrate those who might be socially awkward, eccentric, or cognitively atypical (see box 2.1). Moreover, because networked individualism is premised on individual mobility, it suits well-off, itinerant professionals better than those who are not so physically mobile, namely children, disabled persons, and the elderly.[11] The latter are put at a disadvantage in contemporary urban spaces defined by transportation networks premised on individual ownership of a private automobile. Simply put, those unable to drive to visit friends and loved ones must see them less often.

For others the problem with networked individualism is that it is not experienced as a satisfactory form of community. For example, Alex Marshall, a person helping to build New York City's first cohousing project, believes "that the generally fragmented lives so many of us lead break up marriages, disturb childhoods, isolate people when they most need help, and make life not as much fun. We live, to speak frankly, in one of the loneliest societies on Earth."[12] Similarly, some stay-at-home mothers blog about the loneliness of contemporary suburban motherhood, despite the opportunities for social connection afforded by digital devices. Indeed, one described going to Target just to be around other people but ultimately lacking the courage to ask a fellow parent-shopper: "Are you lonely too? ... Can we be friends? Am I freaking you out? I don't care. HOLD ME."[13] Loneliness among new stay-at-home mothers is not solely a byproduct of suburbia or the Internet age, of course; nevertheless, their plight illustrates how some remain poorly served by networked social life. Such sentiments—as well as the existence of social movements striving to build more communal urban spaces and establish food cooperatives and farmers' markets—demonstrate that not everyone is content with the kinds of togetherness offered by suburbs, supermarkets, and social networks.

Box 2.1

Thick Community and the Integration of Atypical Members

Donald Triplett was the first person in the United States diagnosed with autism. In his early years he was significantly challenged in terms of his functioning, unable to feed himself and eventually unwilling to eat, seemingly oblivious to environmental dangers, prone to violent tantrums at the slightest deviation from his routine, and mostly nonverbal. After institutionalization proved unhelpful, if not damaging, Donald's mother took a far different tack: ensuring that he was included in the community, engaged in meaningful activities, and educated. While growing up in Forest, Mississippi, working on a family friend's farm, and attending public school, Donald's oddities were accepted and his strengths recognized. Classmates and others saw it as their duty to help Donald integrate and improve his abilities, such as by trying to help him to learn to swim or use slang. Someone witnessing Donald as an adult would have scarcely been able to imagine his rough early years: he graduated from college, traveled solo around the world, learned to drive and play golf, and—most importantly—became part of the social life of Forest. Indeed, when researching Donald Triplett for their history of autism, John Donvan and Caren Zucker experienced one manifestation of the community's caring for him: being warned on several occasions, "If what you're doing hurts Don, I know where to find you."[15]

Even if some traditional communities might have been just as likely to excommunicate atypical members and although contemporary social networks no doubt help people with autism connect more easily with others like them, the case of Donald Triplett nonetheless illustrates something special about the practices of integration that are possible within thick communities. As Donvan and Zucker note, "being accepted, even embraced, by the community ... supported a fulfilled life [for Donald], with a network of people watching out for him."[16] That others watched out for Donald did not seem to be because of his ability to "entrepreneurially operate" networks of social ties or because he had really all that much to offer in the networked marketplace of social exchanges; indeed, his ability to hold a conversation has remained limited his entire life. Rather, citizens of Forest appear to have cared for Donald because he was part of the community and that is what members of their community do for one another.

Despite the hype about virtual community, some people do not find computers or smart phones to be suitable gateways to communal belonging. Recent research has uncovered the existence of "cyberasocials," people unable to feel a sense of social connection through digital devices.[14] For such people, a Facebook message or a friendly text simply does not register. Their need for social connection depends on physical copresence. As a result, the increased digital mediation of social connection amounts to a form of discrimination: it specifically diminishes cyberasocials' opportunity to experience belonging.

The Naturalization of Networked Individualism

Networked individualism is political—a social change entailing winners and losers. Yet this political side of networked individualism is rarely dealt with in a substantive way. There seems to be a reluctance to recognize the networked individualization of community as a sociopolitical process. Too little attention gets paid to the fact that the *how* inexorably shapes the answer to *who gets what*. The range of possible ends is influenced by the available means, legislating outcomes just as well as laws do. The lack of curb cuts or entrance ramps, for example, limits access by the disabled. Expensive automated machinery can facilitate the centralization of an industry just as well as lax antitrust regulations.[17] Likewise, technologies help determine which citizens enjoy what kind of social life. Although walking to a neighborhood pub or café remains ostensibly legal in America, it is technically prohibited: the spatial organization of most suburban neighborhoods and prevailing zoning codes prevent such places from existing in the first place.[18]

Technologies are not political solely in terms of prohibiting or discouraging certain actions: They collectively act as "forms of life" as well. Artifacts, techniques, and systems "generate patterns of activities and expectations that soon become 'second nature.'"[19] As the activities they support become taken-for-granted habits, the very conditions of everyday life often change dramatically. Consider how asking for directions today as often as not results in a quizzical look and the question "Don't you own a GPS?" Similarly, a television is not simply an isolated entertainment device, for we live in a world where television cannot really be turned off. The ubiquity of TV has led it to seep into nearly every corner of day-to-day life, from surrogate babysitting to a focus for water cooler conversation. Even those who do not watch TV must contend with a culture shaped by television and its ancillary practices.

In this vein sociologist Ray Oldenburg criticized networked theories of community as being merely the best fit for "the disastrous spatial organization of the typical American city."[20] That is, a networked understanding of social life only begins to make sense in a world where suburbia has become as enmeshed in culture and in the minds of citizens as it has in the physical landscape.

Social network theorists, to be fair, do recognize that the Internet has *affordances* for the practice of networked individualism.[21] The design features of much of today's infrastructure correspond well with the practice of maintaining fragmented social ties. Not enough attention is given, however, to how these affordances (and corresponding constraints on thick community) have political consequences. The end result of mass suburbanization is not simply that citizens more often pick up the telephone, get in their car, or go online to connect with loved ones but that neighboring, frequenting the local café or pub, and pedestrianism are, practically speaking, rendered less viable. Technological changes lead to certain ways of life becoming harder to realize. Again, altering the how influences who gets what.

This insufficient attention to technological politics can be partly traced to a misunderstanding—visible in some network scholarship—of the dynamics of large-scale sociotechnical change. For instance, Rainie and Wellman assert:

> The impact of technology unfolds in three stages. The first stage is substitution as new technology performs older technology's tasks more efficiently. The second stage is enlargement as new technology is used to increase the volume and complexity of tasks that old technology used to perform. The final state is reconfiguration as new technology fundamentally changes the nature of the things it was created to address.[22]

The idea that technology unfolds according to a simple logic of increasing efficiency and complexity, however, has been undermined by decades of technology studies research. It is simply too easy to refute. The introduction of pneumatic molding machines into one nineteenth-century reaper factory actually produced an inferior product at higher costs.[23] The machine's purpose, in contrast to commonly held myths about automation, had less to do with increasing efficiency than with eliminating the skilled workers who had organized the local union. Similarly, while cars can be the most time-effective means of travel in low-density areas, they are grossly inefficient with respect to energy and resource usage. The dominance of the automobile in North America has as much to do with sociopolitical factors as with the technology's perceived advantages. Companies like General Motors, Firestone, and Standard Oil worked together to buy out electric trolley companies and replace streetcars with buses. The unattractiveness of buses to transit users precipitated declines in ridership, and lower profit margins led to further service reductions. The situation was exacerbated by municipal requirements that trolley services extend their lines into low-density suburbs without financial assistance, as well as by the Great Depression, which saddled transit companies with debt.[24] Finally, highway building and suburbanization have been heavily subsidized by governments. If end-consumers had been required to pay more of the costs, the automobile would have been much less appealing.

Much of the bigger picture is lost when technological changes are explained solely in terms of efficiency and similar concepts; the embedded political interests and values are swept to the side. Efficiency may be influential for a narrow range of technical issues, such as when a novel engine design improves vehicle mileage, but most cases are far too complex for the concept to be very helpful. At worst, it is a post-hoc rationalization or makes sense only once the values built into the technology become taken for granted, as in the widespread assumption that moving two tons of steel at high speeds is the best way to transport a single individual to work in the morning.

Ignoring the politics of sociotechnical change implicitly naturalizes it. Rainie and Wellman, for instance, advised, "Technology continues to spread through populations, so the emerging need is for people to learn how to cultivate their networks—and to get out from the cocoon of their bounded groups."[25] "Technology," however, is not a singular thing that spreads as if by its own accord but is a linguistic category used to describe the stuff that humans build. It is far better to talk about technolog*ies*, which are developed, advocated, and embraced *by some people* for various reasons, including gaining an advantage over others. Sociologist Eric Klinenberg likewise argued that it is better simply to accept networked individualistic solo dwelling as "a fundamental feature of modern societies" and dismissed doing otherwise as "indulging the social reformer's fantasy."[26] Such depictions of technological evolution risk repeating the mistakes of mid-twentieth-century "modernization theory" by giving the impression that there is some universal social development pattern inherent to becoming "modern." Although such claims may simply reflect a degree of pessimism regarding the feasibility of realizing alternative modernities, they can imply that networked individualism is the natural or unavoidable state of technological civilization.

This process of naturalization limits the possibilities for imagining or striving toward alternatives. If democracy is the ability for citizens to collectively govern the structures that shape their lives, and if technologies are a subset of such structures, then depoliticized technological discourse is fundamentally antidemocratic.[27] It frames innovation as outside the purview of collective debate and control, as if sociotechnical change were governed by some logic internal to *technology*, not by people. Such discourse is hardly new. Consider the motto of the 1933 World's Fair: "Science Finds, Industry Applies, Man Conforms."

Depoliticized technological discourse, moreover, can become a vicious circle. Technological development often *appears* autonomous because citizens have collectively failed to exert adequate control over it—in part because more sophisticated mechanisms of governance have yet to be implemented. When technological drift is interpreted as autonomous, the steering of technological development is more rarely attempted. Hence, societies fall victim to *technological somnambulism*, largely sleepwalking through the process of large-scale sociotechnical change.[28] Humanity's half-blind fumbling from one sociotechnical change to another begins to be confused with technological evolution. It starts to be unimaginable to subject new gadgets or technological systems to scrutiny *before* they become widely entrenched. The interstate highway system, for example, was mandated by the U.S. government in 1956, while much of the public debate concerning its potential consequences for community life came later. A transportation plan not amounting to a massive governmental subsidization of suburbia could have unfolded if enough people had been empowered to raise critical questions much earlier. To be fair, given most people's lack of experience with governing technological development, the mistaken belief that technological development progresses according to an autonomous logic is in some respects entirely understandable, though politically disabling.

At the same time, even advocates of thick community could better deconstruct the politics of entrenched technologies. For instance, Robert Putnam ended *Bowling Alone* with a call for the development of communal alternatives to television and the building of neighborhoods more compatible with shorter commutes and with public sociability.[29] Given the scale at which television and other screens are embedded in everyday life as well as the sheer momentum of suburban building forms, it would not be surprising if many of his readers remained pessimistic. Without careful attention to the sociopolitical factors sustaining the status quo—for example, the momentum of suburbia as resulting from massive levels of subsidization, the routinization of major repairs and restorations, and numerous other entrenched practices and beliefs—the present can easily appear "natural" or impossible to change. Many too quickly forget that the hollowing out of most major North American downtowns during the expansion of suburbia in the second half of the twentieth century was a colossal political undertaking, costing billions or even trillions of dollars in wasted urban infrastructure as resources were redirected to the suburbs. It took the decisions and failures to act by innumerable business executives, urban planners, local politicians, and others to nearly decimate America's urban centers.

Reverse Adaptation as a Partisan Position

Innovation could be guided to preserve or to enhance a variety of forms of community life. Even those who see a lot of good in contemporary networked individualism might want to make room for more debate than is now occurring. Otherwise, they risk giving tacit approval to *reverse adaptation*: the process in which human ends are molded to match the available technical means rather than the other way around, even if the original ends are distorted or replaced with something very different.[30] Reverse adaptation has too often been the standard answer to concerns about communal decline. Consider Herbert Gans's suggestion, more than a generation ago, that widespread feelings of social isolation in the emerging suburbs might be best dealt with not by pursuing design changes to encourage thicker forms of local social connection but by lowering the cost of telephone calls.[31] Even eminent scholars get trapped in the status quo.

Reverse adaptation to networked individualism looks unequivocally desirable only after brushing aside all the dimensions of community life apart from the existence of social networks, including strong practices of civic engagement and reciprocity; material interdependence and shared risk; and a felt psychological attachment to places, institutions, and groups. Community has traditionally meant a more coherent and longer lasting social entity rather than merely the plural of personal friendship. Because the *theory* of networked community leaves these aspects of belonging out, it too easily motivates reverse adaptation to the *phenomenon* of networked individualism. Sociologist Ray Oldenburg went so far as to argue that network theory "perverts the concept" of community and mainly sustains it as a myth in the face of increasing atomization.[32]

Indeed, networked individualism as *both a theory and a phenomenon* seems built on a narrow understanding of social relationships. Internet studies scholar Michele Willson, for instance, has critiqued social network theorists for starting with the presumption that humans are atomistic consumers of social connection.[33] Doing so privileges individualistic and instrumental understandings of social relationships, something that is clearly visible in sociologist Eric Klinenberg's study of the networked individualism of people choosing to live alone. His interviewees championed a mode of sociality by which social ties are kept at arm's length. They lived by themselves and eschewed strongly binding social obligations to avoid the inevitable conflicts, disagreements, and compromises that come with living with others. The best thing about living alone is "knowing that I don't have to consider anyone else," reported one woman in the study.[34] A young man preferred to mostly keep to himself and cultivate self-reliance in order to "get used to it when still young," because all relationships are destined to end in tragedy or separation.[35]

In previous studies of community, the attitudes espoused by Klinenberg's interviewees would have evoked concern. Sociologists such as Robert Bellah would have described them as steeped in a "language of separation" and threatening the stability of committed relationships and civic life.[36] What sort of social life can exist if citizens come to view committed bonds as only burdensome and anxiety provoking? Moreover, such approaches to social life cannot be taken at face value, as evidence of the need for network individualism, because a preference for minimizing social bonds is not simply a matter of individual choice but a product of the context in which one is born. Growing up without thick community shapes people's preferences for solo dwelling, among other practices. While Klinenberg's interviewees no doubt have the right to "choose" the type of social life they want, there exists a real risk of valorizing such choices to the extent that networked individualism becomes presented as a model by which everyone should live.

Network advocates' scholarly arguments would be strengthened by serious engagement with such concerns. Unfortunately, Klinenberg, among others, have dismissed the worry that networked individualism contributes to the decline of civil society on the grounds that "its vague generalities distract us," arguing that we should focus instead on providing better support to the "truly isolated" and on recognizing the positive features of solo dwelling.[37] It is perfectly natural, even desirable, for scholars to have different priorities regarding research. It is difficult to understand, however, why any social scientist would outright reject inquiry into actual and potential qualitative changes in community. One possibility is that value commitments may be embedded in the scholarship, perhaps so deeply that even the authors lack an awareness of it.

The portrayal of the networked present as continuous with the past also gives tacit acceptance to reverse adaptation. Recall Keith Hampton's equation of Facebook status updates with the community work previously done on front porches.[38] Such discourse seems like a continuity argument: an exaggeration of the resemblance between an emerging practice and past activities, which results in novel risks or consequences being minimized or ignored. As technology ethicists have argued, a continuity argument "is often an immunization strategy, with which people want to shield themselves from criticism and to prevent an extensive debate on the pros and cons of technological innovations."[39] Barry Wellman, for instance, has used the observation that networked individualism predates the Internet in order to dismiss worries about accelerating networked individualization.[40] Indeed, he has insisted that those concerned about communal decline suffer from "misplaced nostalgia," asserting that "researchers have found thriving communities wherever they have looked."[41] On the one hand, Wellman may be simply trying to counter the pessimism or doom saying that often infects cultural commentary regarding technological change. On the other hand, in overstating his case, he ends up implying that the continuity of friendship networks is all that should matter with respect to social life. Such a position undermines the possibilities for debate about potential declines to other dimensions of community.

While the above arguments might seem pragmatic, given the apparent momentum of network technologies, they implicitly serve the interest of some while treating those not benefitting as if they did not exist. To advise citizens to simply forgo their bounded groups to better cultivate their social networks[42] is to forget that some may be harmed by the decline of bounded groups. Similarly, recall Zeynep Tufekci's research uncovering "cyberasocials," people whose need for a sense of social connection cannot be provided through digital devices. She nevertheless concluded, "We would be better off debating how we can use new communications technologies to combat the economic, political, and cultural forces that threaten to tear us apart" rather than inquire into Internet technologies' role in social isolation.[43] Taking new communications technologies in their present form as a given seems practical, insofar as the barriers to substantially redesigning them remain large. Such a move, however, ends up sidestepping the question of what to do about cyberasocials. Any expansion in the use of digital technologies likely serves them poorly.

There are still other possible explanations for the tendency of scholarly discourse around networked individualism to slide toward reverse adaptation. Social scientists' interpretations of sociotechnical change are likely influenced by their life experiences as itinerant professionals. Given that the idea that people's views and perceptions are shaped by their social roles is foundational to the social sciences, it would be odd if they were not so influenced. As sociologist Stephen Brint has noted, "social scientists ... live in a world in which a great variety of social ties are juggled and the pursuit of valuable connections looms large both at work and in informal social settings. This may encourage a view of social relations as intense but fleeting and as significant primarily for the instrumental benefits that may eventually accrue from them."[44] Because social scientists' life experiences tend not to include being a stay-at-home mother or working in jobs that do not come with career-based networked communities, they are especially at risk of failing to give due diligence to other people's perceptions of communal decline.

A favorable view of networked individualism, moreover, reflects the prevailing moral bias within Western and Westernizing technological societies: liberalism. By this I do not mean the political liberalism associated with the Democratic Party in the United States or liberal parties across the world but philosophical liberalism, a particular moral way of viewing the world with roots in seventeenth-century political thinkers such as John Locke. Philosophical liberalism presumes that the good life is based in the protection and valorization of individual choice.[45] For right-wing liberals, such choices are ideally provided by the "free" market; for left-wing liberals, they are attained through the unrestricted expression of personal identity.

As philosophical communitarians point out, turning choice into a sacred right, an end unto itself, can lead to undesirable consequences. Political philosopher Michael Sandel argued that liberal philosophy would demand that a person be able to choose his or her social bonds autonomously, without coercion. Doing so, however, would require perpetually keeping one's relationships at arms' length, because sustaining relational commitments and strong emotional bonds would impinge on the ability to choose rationally. In Sandel's words, such a person would strive to remain an "unencumbered self," constantly wavering between detachment and entanglement, unwilling or unable to let ties with others become a strongly constitutive part of his or her own being. Sandel argued that such a person, in lacking the capacity for strong commitments, would be without moral depth.[46]

In any case, much of the discourse concerning networked individualism appears to assume a philosophically liberal outlook. Indeed, Rainie and Wellman seemed to celebrate *the choices* offered by networks: "People have more freedom to tailor their interactions. They have increased opportunities about where—and with whom—to connect." Wellman, moreover, first called networked social activity "community liberated," a value-laden label that takes for granted philosophical liberalism's equation of liberty with individual choice and mobility. Klinenberg likewise has described solo dwellers as having grown to "appreciate the virtues of living lightly, without obligations."[47] The avoidance or weakening of binding social obligations, however, can only appear to be an unequivocal freedom-enhancing virtue if one has already embraced a narrowly liberal view of the good life.

The networked individualization of community, contrary to such views, has introduced new constraints and compulsions—not just freedoms. As scholars Williamson, Imbroscio, and Alperovitz noted in regard to the thinning of economic community in the face of neoliberal globalization, "If the hometown where you grew up dies out or decays economically and you are *compelled* to leave, a very significant life option has been extinguished."[48] In such a situation, one's freedoms are reduced, despite the existence of new networks, like Fixxer, Lyft, and AirBnB, that could be used to eke out an individual living. A strictly liberal politico-ethical framework thus renders less visible the broader constraints on modes of life enacted via networked individualization by privileging the expansion of individual choice. Much in the same way that Henry Ford quipped that consumers could have his Model T in any color as long as it was black, networked community offers citizens any kind of freedom they could desire as long as it is a form of liberal individualism.

Finally, is there perhaps a sense in which reverse adaptation represents—ironically—a *conservative* approach to technology and community? Networks are hardly radical any longer, having become the status quo form of social organization in more affluent nations over the last decades, if not century. Hence, to argue that people should only look to better adapt themselves to networked sociality is to advocate for the *conservation* of a way of life that happens to be increasingly imposed on everyone, despite the objections of those who find it less desirable or fulfilling. More often realizing thick forms of togetherness, on the other hand, would require progressive changes to regulations, infrastructures, and institutions regarding technology that would allow for a greater degree of democratic control, rather than a passive resignation to be governed by the policies and technologies—the sociotechnical tradition—one happens to have inherited.

In contrast to far too many books on the question of technology and community, I do not pretend that what follows is a nonpartisan project. My intention is to provide social analysis that can be useful to people who currently wish for a more communitarian technological civilization as well as to those who might seek change if they were pushed to consider the issue critically. It should be noted, however, that *thoughtful partisanship does not imply carelessness with data*. Rather, as the eminent political scientist Charles Lindblom argued, it means that one merely acknowledges and approaches social analyses as inevitably in service to *some* values, groups, and interests rather than others.[49] Otherwise rigorous social scientific scholarship should not falsely give the impression that it serves the interests of everyone equally. Unlike many other commentators on contemporary life, I do not use the word "we" carelessly—if at all.

The next chapter provides the foundation for an alternative way of looking at community, one that recognizes it as a multifaceted and diverse set of phenomena. This reconceptualization of belonging offers a starting point for better examining the communitarian limits of contemporary technologies. That is, I aim to break away from stale arguments that appear to presume that humanity is faced with a dichotomous choice between some romantic ideal of bucolic rural togetherness and mass urban society, or even some melding of the two. Rather than reject practices such as network individualism or virtual community as inauthentic, I depict them as relatively thin instantiations of community among a diversity of possibilities. Moreover, there are practical advantages gained by viewing different manifestations of togetherness as lying on a spectrum with respect to several different dimensions of communality. Community advocates are better off if they can move beyond a general sense of malaise to zero in on the dimensions of communality that they find to be too "thin" or missing in their lives. Any effort to enhance community life is likely to flounder without a clear sense of exactly what one is hoping to strengthen or create.

3   From Thick to Thin: The Seven Dimensions of Communality

Few words are as frequently uttered as imprecisely as the word "community." It is used to label everything from daily life in a rural small town to participation in an online forum as well as groups of people that merely happen to share some ethnic, racial, or sexual identifier (e.g., the gay community). These cases are clearly not communal in the same way. Small town residents economically depend on one another for their livelihoods. Their small talk and gossip is as much for cementing social norms and doing politics as for maintaining social bonds. Online forum users or weekly "meetup" participants, on the other hand, tend to be seeking some very specific good, not just social connection. Language learners meet up in bars and cafés in order to acquire speaking practice as much as belonging. The modicum of connection felt by scanning one's Facebook newsfeed may partly allay a sense of loneliness but hardly amounts to anything like what community has historically entailed. Members of the "black community," in contrast, are bound together not by social ties or material interdependence but by a shared culturally assigned identity—including the experiences of racial discrimination and oppression that typically come with it. The tendency to refer to each of these very different phenomena with the same term—without qualification—obstructs clear thinking about sociality.

The core of community is the provision of social belonging, which social psychological research has demonstrated to be a fundamental human need.[1] Its polar opposite, loneliness or social isolation, causes intense psychological and physiological harm. Like other aspects of well-being, however, exactly how and in what way belonging should be provided is an intensely divisive issue. The need for togetherness can be met more or less well through a number of different mechanisms, each coming with its own implications for social life. The sheer diversity of often mutually incompatible communal forms and the complexity of belonging as a social phenomenon make universalizing definitions problematic. Even worse, there is a tendency to reduce community to a single variable and deny this diversity. How could the concept of community be put to work in a way that better acknowledges the complexity and diversity of social life?

This chapter attempts to untangle social belonging from this theoretical morass by conceptualizing it as a multidimensional social phenomenon. Some instances of community can be described as *thicker* than others in terms of their strength in each of the possible dimensions of communality as well as with respect to the number of dimensions actually present. An online forum, for example, provides relatively thin networks of social ties, which at their best involve moderately thick relational exchanges of social and material support. Their economic and political dimensions are typically thin, however, if not nonexistent. Users are rarely tied together by material interdependencies or involved in the governance of the site. In any case, this framework is both broad enough to encompass most forms of community, including networked individualism, and systematic enough to distinguish between them. As a result, it moves the debate past dichotomous notions such as "authentic community" toward a consideration of different weightings of the various dimensions of communality. My goal here is to provide a framework that allows for the improved differentiation of the myriad forms and constructed understandings of belonging that exist today.

Community as Webs of Social Ties

In most accounts, community is made up of webs of social bonds. However, these ties can have a variety of different topologies. Those who study more traditional forms of community life describe it as built on dense, multiplex, and systematic social ties.[2] Bonds are dense to the degree that every possible connection between two community members actually exists. Do members know most of their compatriots either directly or through their other relationships? Multiplexity describes the extent to which members' social connections are integrated into multiple spheres of everyday life. Bonds are multiplex when they are not functionally segregated. In a cohesive neighborhood or rural village, residents tend to see their neighbors and friends in multiple contexts: at work, church or temple, the local pub, and/or civic association; when shopping or during their volunteer time; and when engaged in political action. Ties are systematic when they are ordered and incorporated into larger social bodies. For instance, so-called traditional, or tribal, societies tend to be highly systematic. One belongs to a family within a clan or band that fits into a larger tribe and so on.

The opposite end of the spectrum, acommunality, has been typically defined within sociology as characterized by diffuse webs of social ties that are more transitory, contractually defined, and fragmented than in traditional communities. Sociologists from Ferdinand Tönnies to Georg Simmel associated the decline of communal social webs with the greater anonymity, mobility, and individualism within the then-emerging industrial metropolises of the late nineteenth and early twentieth centuries. The social ties of the typical urban dweller were viewed as less intimate, more superficial, and more likely to be segregated to only one sphere of everyday life. The acommunal urbanite was seen as flitting around a network of weak associations, acquaintances, and contractually governed economic and bureaucratic relationships rather than being enveloped within a strong web of supportive, meaning-giving ties. Some sociologists, Simmel among others, wrote positively of the metropolis for giving birth to new forms of individual freedom at the same time that they lamented the development of blasé and apathetic attitudes among urban dwellers.[3]

Despite the anti-urban slant of early sociologists, webs of dense, multiplex, and systematic social ties have been found in large cities, though often in ethnic enclaves. Urban sociologist Herbert Gans, for instance, described the vibrant community life and dense webs of ties that enabled Italian Americans in Boston's West End to navigate life in an urban ghetto. Similar levels of communality have been found among first-generation Irish immigrants to London.[4] In other ways, however, ethnic enclave communities are less than ideally dense, interwoven, and systematic. Gans described West End residents as feeling alienated from their work lives in factories outside the neighborhood, speculating that they might have compensated for their lack of fulfillment by focusing on their relationships with kin and neighbors. In any case, Gans contended that the social aloofness that scholars commonly ascribed to city dwellers really only fit a particular population: the bohemian or well-to-do cosmopolite.[5] Dense, multiplex, and systematic social ties can form nearly as well in urban spaces as in rural towns and hunter-gatherer bands. Indeed, the aim of new urbanist design movements, as discussed in the next chapter, is to mold urban form to be more compatible with such social arrangements.

Others, in contrast, have maintained that the association of community with dense webs of social bonds confounds it with feelings of local solidarity. Network sociologist Barry Wellman has argued that because most people's primary social ties are located no longer in dense, tightly bound, and localized groups but in diffuse social networks, it makes little sense to equate the former with community.[6] In his perspective, community is a matter of individuals being connected via intimate and weak ties to social support and important resources, not the existence of a feeling of connectedness to a place or to local webs of mutual obligation. As such, the topology of social ties is rendered mostly irrelevant to the question of whether community exists. Rather, it is considered to occur wherever there are social networks, whether these networks take place within a neighborhood, an online forum, or across a nation. From the perspective of scholars who associate the term with webs of dense and multiplex ties, however, Wellman describes not community but friendship networks.

Rather than either follow or dismiss Wellman, it is better to think of diffuse network ties as lying toward the end of a spectrum of possibilities, while on the opposite end are forms of social connection rooted in dense, multiplex, and systematic bonds. The former is a *thinner* arrangement of communal ties than the latter, which is more characteristic of *thick community*. Both can provide belonging in a real sense but in very different ways. Most communities, of course, probably lie somewhere in the middle, being composed of some mixture of fragmented networks and dense social webs. Regardless, whether communal bonds are diffuse or dense affects the character of relational practices, understandings of the self and community, as well as economic and political interactions between members. In the next sections, I explore these other dimensions in greater detail.

Community as Relational Exchanges and Social Support

Community is more than the mere existence of social bonds. The practices between those bonds also matter. Communal norms and practices are typically distinguished from the realm of contractual economic exchange. Psychologists define relationships as communal by the extent to which participants care for one another's needs without expecting immediate reciprocal benefit.[7] Exchange relationships, in contrast, are rooted in the specific and more immediate exchange of similarly valued goods, as is the case within markets. In communal relationships, considerable effort is spent to ensure that reciprocity is construed as nonspecific and nonimmediate. Gift givers usually remove price tags in order not to convey a sense of specific debt or obligation. Similarly, taking someone out to dinner is reciprocated only at a later time and usually not at the same restaurant, if reciprocated via a meal at all. Of course, it is not only material aid that is provided but also caring in times of need, attentive listening, and companionship.

Sociologists make a similar distinction when differentiating generalized and specific reciprocity. Robert Putnam has associated the former with "social capital": the existence of high levels of trust, interaction, and cooperation that recursively support community life.[8] Generalized reciprocity is characterized by the act of giving or caring for others without expectation that they immediately return the favor. Rather, it is done because giving aid is the norm and one believes that the favor will eventually be returned in the future—though not necessarily by the same person. Acts of generalized reciprocity can be as insignificant as holding the door for someone or as large as joining with neighbors to raise a barn. Moreover, as former Missoula mayor Daniel Kemmis depicted in his account of rural Montana barn raisings, strong mutual aid norms even elicit the help and cooperation of neighbors who may not even like each other.[9] In some cultures, practices of mutual aid are separated from gift giving: recipients are discouraged from even expressing gratitude because it could imply that providing aid was a voluntary choice and that a debt was incurred. Inuit walrus hunters, for example, would share their catch with the insistence that "since we are human we help each other ... What I get today you may get tomorrow."[10]

In the psychological literature, such actions are understood through the label *prosocial behavior*: actions directed toward the benefit of others and not motivated by egoistic gain. Although prosocial behaviors such as stopping to help a stranded motorist may be prevalent even in very thin communities, social psychologists have explicitly connected it with thick community.[11] That is, a propensity for prosocial actions is related to the degree to which people are integrated into a network of communal bonds and the stability of those bonds. Furthermore, prosocial action forms a positive feedback loop with a felt sense of belonging and attachment to a psychological or geographic community. For example, volunteering at a nearby hospital or local HIV/AIDS service organization establishes and maintains social bonds and fosters a sense of belonging, which, in turn, inspires more volunteering. In line with Putnam's theory of social capital, prosocial activity both drives and is driven by dense social connections and the feelings of social trust stemming from a sense of community.

Many everyday practices of social and material support, of course, lie between specific exchange and mutual aid. Material exchanges within communities often differ significantly from the ideal models of economic theory. Community members are interested not simply in making a living from their productive activity but in sustaining relationships too. For instance, early promoters of home photovoltaics and small-scale wind or hydropower plants eschewed traditional business models.[12] They included their customers as cooperative participants in the process of building and design rather than as passive consumers, and their employees more or less set their own hours. These business owners focused less on maximizing profit and more on ensuring the spread of renewable energy technologies in their communities. Similarly, local businesses often cut into their margins when working with nearby institutions. One Troy, New York, church, for instance, maintains an ongoing economic relationship with a local firm that goes beyond mere contractual obligations; the company managing church property performs maintenance work at a substantially discounted hourly rate, reflecting a sense of social obligation and a prioritization of the stability of the arrangement. Economic exchanges are communal to the extent that broader social and publicly minded goals shape the conduct of business, not just short-term profit maximization.

Likewise, the more specifically reciprocal gift exchanges dominating rural village life in non-Western nations do not create wealth so much as they create long-lasting relationships of credit and debt that bind individuals and families together. Anthropologist David Graeber has described how women in a Nigerian village would walk considerable distances to give a handful of okra. All gifts had to be reciprocated so that one did not appear to be a parasite or an exploiter. Moreover, the women intentionally did not match the perceived value of the previous gift in order to spur future giving. Although economically irrational, such activity makes sense when understood socially, because "in doing so, they were continually creating their society."[13] Frequent exchanges and alternating obligations sustain the interdependencies upon which thickly communal social arrangements are built.

Not only are economic exchanges often governed by communal logics, but otherwise communal relationships can be framed in terms of market-like exchanges. Sociologist Robert Bellah and his coauthors characterized this relational practice and disposition as "therapeutic contractualism," a way of modeling and negotiating social bonds as if they were contractual exchanges.[14] Relationships, under these terms, are only important in terms of the utilitarian or expressive benefits accruing to individuals—an egocentric way of understanding the social world that Bellah et al. associate with psychotherapy. Sociologist Eric Klinenberg, for instance, displayed a therapeutic understanding of marriage when framing it primarily as a means to access those "certain things you can only learn about yourself [when] living intimately with another person."[15] Under a contractual model, moreover, relationships are always tentative arrangements, maintained insofar as both parties keep their side of the bargain. If a friendship or marriage ever seems more costly than the social support or good it provides to the self, participants are to dissolve it; maintaining relationships for any other reason is framed as illogical.

Although social network theorists do not make the connection, therapeutic contractualism appears to be the dispositional logic undergirding and reinforced by the practices of networked individualism. As such scholars already recognize, "personal networks rarely operate as [communal] solidarities."[16] Networked reciprocity is generally "tit-for-tit" rather than "tit-for-tat." That is, it involves the exchange of material or social support with specific people who generally return the same kind of good. This is not surprising: generalized reciprocity would make little sense in fragmented and transitory social networks. There is no larger sense of "we," no long-lasting forms of interdependence that would motivate spontaneous mutual aid.

Although therapeutic contractualism is probably communal enough for some people much of the time, it has its limits. A recent description by writer Gina Tron of the lack of support provided by her social network after her rape illustrates the *thinness* of contractual logics.[17] Fearing emotional outbursts or not wanting to hear a retelling of the incident, members of Tron's personal network purposefully avoided her. There is no guarantee, of course, that less therapeutically contractual ties would always be more supportive in such cases. Nevertheless, the contractual view appears to dispose people to fail to provide care whenever specific reciprocity is unlikely or impossible.

Talk

The most important communal exchange occurs through talking. As biolinguist John Locke has argued, speech is too often mistaken as merely a vehicle for information. Even seemingly unimportant small talk and idle gossip help cement social bonds, facilitate the informal negotiation of shared norms, build trust, and lessen feelings of loneliness. Face-to-face talking is "social grooming," a practice similar to the physical grooming utilized by humanity's ape cousins to foster group cohesion and lessen anxiety.[18] Furthermore, as Janet Flammang has argued, the kind of talk undergirding local civil society begins at the dinner table. The enjoyment of conversation around shared meals helps instill the baseline affective bonds and linguistic skills citizens need in order to sustain relationships through inevitable moments of conflict.[19]

The maintenance of places for talking, unsurprisingly, has been a major component of social life. Indeed, sociologist Ray Oldenburg well described the particular amenability of local cafés, pubs, and the German Biergärten to relaxed forms of talk and their importance for social bonding.[20] Likewise, communal wells have traditionally provided local people the opportunity to gossip and build relationships as they gathered water and did laundry.[21] Community is destabilized by the loss of such places, even in less traditional societies: some have suggested that the decline in civility and fraternity between conservative and liberal congressional legislators in the United States is due partly to the fact that they more seldom dine with one another after work hours.[22]

Given oral communication's role in developing and stabilizing social bonds, the thickness of any community can be characterized by the frequency and intimacy of face-to-face talking. Indeed, even members of otherwise relatively thin communities, like academic research specialty areas or anime fans, will often insist on meeting in the flesh. Doing so helps cement social bonds in an affectively rich way. Even the foremost champion of virtual communities, Howard Rheingold, pointed to the frequency and intimacy of offline social gatherings to justify the authenticity of the togetherness he pursued on an online forum.[23]

Communication via non-face-to-face mediums, of course, likely provides some of the same benefits as embodied interaction, albeit less reliably across circumstances and populations. Recall how cyberasocials do not feel a sense of social presence via digitally mediated technologies.[24] Moreover, anyone who has been in a long-distance relationship has probably felt some dissatisfaction with the intimacy available through phones calls and Skype chats. The communicative and affective value of hugs, hand holding, and mere proximity cannot be overstated. Of course, the distance provided by more mediated communication media can occasionally have its advantages, including better enabling a "cool off" period after a fight. At the same time, these media can provide access to social belonging to those excluded by the physical communities that surround them. Nevertheless, the advantages of such media do not offset the relative thinness of the social connection they provide.

The Symbolic and Psychological Sense of Community

Aside from manifesting in practices and social bonds, community exists in the minds of its members. As political historian Benedict Anderson put it, "all communities larger than primordial villages of face-to-face contact

---

Box 3.1
Thick Sociality on the Island of Ikaria

Stamatis Moraitis lived a relaxed but socially active life in Ikaria. Nearing his one hundredth birthday, his days were occupied with working in the garden and maintaining a vineyard that produced four hundred gallons of wine per year, while evenings consisted of visits to the local tavern to play dominoes with friends. He and his neighbors would frequently walk to each other's homes to chat over glasses of wine or cups of tea, and Sundays were usually defined by church-related gatherings. Even if Stamatis had wanted to live an isolated existence, his fellow Ikarians probably would not have let him. Nearly everyone, regardless of age, is cajoled out of their homes to share in feasting and dancing at festival time.

Despite the island's 40 percent unemployment rate and despite having fewer of the material luxuries often associated with "modern" life, people living in Ikaria live longer, healthier, and more relaxed lives than those residing in the world's metropolises and suburban areas. Indeed, afflictions like dementia and depression are far less common among Ikaria's elderly. Social life on Ikaria is characterized by the dense webs of ties and the frequent face-to-face interaction associated with thick community. Although the relative lack of privacy on Ikaria would likely feel overwhelming to those who have been raised to see the physical isolation of suburbia as normal, if not eminently desirable, it helps ensure that residents like Stamatis are always connected to and hearing about fellow community members.[25]

(and perhaps even these) are imagined."[26] That communities are imagined does not mean they are false or inauthentic but rather that a significant part of their felt realness comes from the symbolic or psychological construction of a "we." It matters a great deal, however, which processes and practices are used to evoke or instill a sense of community.

On the thin side of the spectrum lie symbolic forms of belonging almost wholly detached from direct social relationships. Nations are one example. The average American has no hope of knowing—or even knowing about—more than an infinitesimal fraction of his or her 300 hundred million or so compatriots. For them, the United States exists as a collectively held imaginary: they think, talk and act as if it were real. Indeed, this imaginary is real enough for some citizens to be willing to kill or die for it. This felt realness, of course, does not come from nowhere. Anderson connected the rise of nationalism, in part, with the standardization of print languages and the emergence of national media. "America," similarly, is reinforced as an imagined community through stories, symbols and rituals, from the pledge of allegiance to national monuments, moments of patriotic celebration, and sporting events.[27]

Nations are not the only cases of discursively created symbolic community. As social theorist Craig Calhoun has pointed out, "communification" via print and mass media occurs for a range of "categorical identities" (e.g., gender, ethnicity, race, lifestyle, sports team affliation).[28] The "black community," for instance, is not created only through the shared experience of racism but through everyday media, such as *Ebony* magazine, and the political discourse of civil rights leaders and participants as well. Media-driven symbolic communities have their advantages and can be thicker than one might otherwise imagine. A case in point is how gay rights advocates have used sites like Gay.com to help organize offline political action. On the other hand, owners of such media portals are usually in the business of promoting not civic engagement but rather the delivery of a desired demographic to advertisers. Forum members typically lack a substantive say over which stories and discussions are permitted on the site; they are more often consumers than communal participants.[29]

Symbolic community creation also occurs for physical places through moments of "civic communion."[30] These are moments when members rhetorically celebrate their common existence, speaking of the social and material connections between them as well as shared valuations of local ways of life, collective goals, and political structures. Occasions and places for civic communion can include annual festivals and local heritage museums. Such moments, of course, are not always rosy. For instance, political conflict in Manhattan, Kansas, over a Ten Commandments monument placed in front of city hall broke local residents into oppositional "rhetorical communities," which dissolved only when the two sides hammered out a compromise. In any case, place-based symbolic community is built on webs of social ties and interdependencies less present in the case of nations or categorical identities. Moments of rhetorical communion are less and less feasible, however, as networked citizens become habituated to describing their political action and values in terms of individual utilitarian or expressive benefit.[31]

The temptation to reduce belonging to its symbolic dimensions should be avoided. Consider anthropologist Anthony Cohen's argument that community "inheres in [people's] attachment or commitment to a common body of symbols" more so than structural interconnections. He has insisted that the Saami, an indigenous group in Norway, persists as a community as long as it is sustained as a collective mental construct, despite the blurring or elimination of structural boundaries by the increasing encroachment of white European technological civilization.[32] Although this view is correct in an important and real sense, it risks diminishing or denying the value of the other dimensions of communality. At any rate, maintaining a shared commitment to a common body of symbols absent strong facilitating institutions would require heroic efforts.

The reduction of belonging to its symbolic dimension is visible *in practice* wherever a contrived image of community is used to psychologically manipulate citizens as consumers. Indeed, people are increasingly engaged in the consumption of symbols of togetherness: for example, shopping malls, the Olympics, public grieving for dead celebrities.[33] In such cases, citizens are merely bound together through the collective act of buying and viewing rather than by relational or material interdependence. These rituals often feature a simulation of sincerity designed to evoke a sense of belonging, a technique media scholar James Beniger called "pseudo-community."[34] Television shows like *American Idol* and *Britain's Got Talent* frequently employ tactics like having hosts act as if they were conversing directly with viewers and framing the audience as a community bound together by the act of voting for their favorite performer. The purpose of these methods is ultimately not the viewer's well-being, of course, but to boost ratings and advertising revenues.

Similar to the sense of belonging, the image of community-as-place is often contrived and exploited for commercial gain.[35] A case in point is Boston's Faneuil Hall Marketplace. Although publicly owned, it is little like the marketplaces that have traditionally centered community life. It is a dressed-up mall, serving tourists interested in consuming the manufactured image of a community market. Similarly, builders of gated communities sell the resemblance of a quiet, safe, semirural community to affluent whites rather than an actually functioning one. Indeed, one trenchant critique of suburbia, strip malls, and fast food chains is that they cannot operate as authentic places.[36] According to anthropologist Marc Augé many contemporary urban environments are better characterized as "non-places," functioning mainly as way points for anonymous spectator-travelers rather than coconstructing shared identities, relations, histories, and "unformulated rules of living know-how."[37] That such environments facilitate consumption and the movement of individuals far better than gathering and social belonging, however, does not stop their symbolic construction as places from having power.

As political scientist John Freie has outlined, these are far from the only cases of symbolic community being used to exploit the human need to feel connected.[38] Elaborate "town hall" meetings are staged by political candidates to persuade the voting public that they are "one of them," not an elite funded by other elites to maintain some facet of the status quo. Evangelical churches and their associated televised services offer a sense of community as a therapeutic escape from the hardships of everyday life, yet often remain aloof and detached from local civil society. The symbolic community provided in both cases is thin if not insincere, disconnected from underlying webs of relationships and relational practices of material support and is partly a mechanism for generating wealth for elite politicians and celebrity preachers.

Whether contrived or not, community manifests not merely symbolically but also in the minds of members. Psychologists have characterized the sense of community as the feeling that one exists within a "mutually supportive network of relationships" that is dependable and thereby prevents "sustained feelings of loneliness"[39] Among the contributing factors are the feeling of membership in and influence within the community, the sense that one's needs are fulfilled, and the existence of a strong emotional connection with fellow members. A psychological sense of community often develops as members interact with one another frequently in meaningful and satisfying ways. Participants feel not only a sense of agency in the mutual fulfillment of needs but also that others are personally invested in the life of the community. Scholars stress the importance of members being able to conceptualize the community's borders and membership, view their identity as rooted in communal ties, believe in the efficacy of the community, and concern themselves with its future.[40]

A psychological sense of community, like its symbolic dimension, should be evaluated as thick not simply according to the extent that members feel it but also in terms of its rootedness in other dimensions of communality. For instance, even though some studies show a positive correlation between urban sprawl and social trust, a proxy for the sense of community,[41] the privatized structure of suburban life suggests the need for caution when interpreting such findings, for the commuting times, social homogeneity, and fragmentation of work, play, and domestic life characteristic of suburbia are all connected with declines in civic engagement.[42] Hence, a symbolic or psychological sense of suburban community may have more to do with the marketing of such areas as idyllic semirural settlements coupled with residents' shared ethnicity and social class than with actual patterns of interaction, cooperation, and interdependence among neighbors. Trust comes easy when contact is rare and superficial, albeit pleasant, and the neighborhood is designed to convey a sense of bucolic togetherness.

Likewise, online communities should be evaluated in light of the presence of factors like material support and embodied interactions. Kind words from and acts of rhetorical communion by online connections certainly have meaning, but they frequently lack depth. Members of thicker "virtual" communities might provide financial support to another who has been diagnosed with a deadly illness, babysit their children, and meet for chili cook-offs.[43] Even in such admirable cases, however, the political and economic facets of community life, discussed below, remain fairly thin. In sum, the symbolic or psychological sense of community is probably a necessary but far from sufficient condition for thick community.

Communal Economics

Often underemphasized in discussions of community is the role of economic activity and self-provisioning. Communal practices and norms of exchange are premised on the importance of maintaining relationships and mutual aid, which are not to be overshadowed by the pursuit of individual gain. Trade networks among and within traditional societies often involved the exchange of goods that both sides could easily make themselves, serving more to sustain alliances than to create wealth.[44] As anthropologist Richard Wilk has pointed out, people in Western industrial societies continue to sustain relationships through economic activity. Yet, they do so more through shared consumption than via material interdependence: a shared vacation cruise or shopping with friends becomes the focus of relationship-sustaining economic activity in a society where economic goods and services are primarily exchanged between semianonymous individuals.[45]

Narrowly rationalistic economic ways of thinking, on the other hand, can be acommunal if not outright anticommunitarian. As Harvard economist Stephen Marglin has argued, "[mainstream] economics offers us no way of thinking about the human relationships that are the heart and soul of community other than as instrumental to the individual pursuit of happiness."[46] Indeed, it is hard to imagine how practices of mutual aid or generalized reciprocity would be practiced by the self-interested maximizers of individual utility that much of microeconomics takes as a model of humanity. Although more often challenged today, mainstream economics has long understood practices like communal barn raisings as irrational. The provision of mutual aid crowds out potential entrepreneurs who can realize new efficiencies and profits by cutting off the relational fat. Yet, as

Marglin noted, premarket practices are only inefficient or irrational if one thinks their sole purpose is economic production.

Economistic thinking has been found to have detrimental effects on cooperation and a sense of moral obligation to others. Majoring in economics or being prompted to think about money leads to more self-interested, less prosocial behavior; another famous study found that fining parents who were late in picking up their children ended up increasing lateness because parents came to see themselves as paying for a service rather than owing a moral or ethical obligation to childcare providers.[47] Market thinking can crowd out other social or communal norms.

In contrast to the practices of self-provision, mutual aid, and reciprocal gift giving within communal societies, most economic activity in affluent, Western nations is defined by individualism and impersonality. Although a department store clerk may be pleasant and helpful, he or she is usually a stranger whose interests only temporarily and imperfectly align with the customer's. Indeed, apathy or even antagonism often underlie "ideal" market activity. Consider the controversy over a smartphone application used to scan barcodes in physical stores in order to compare prices with Amazon. The online retailer's success in undercutting local businesses is aided by the extent to which citizens have internalized the economistic thinking that makes such behavior appear to be socially acceptable. A case in point is how one online commentator supported the app by saying, "If you think for one moment that any of the stores you choose to shop at care about your budget or the costs your family incurs each month, you're wrong. They are in business to make money. I'm in the business of keeping my money, and charity starts at home—my family comes first. Not theirs."[48] Such logic makes sense when customers and business owners are understood as no more than independent social atoms competing over scarce dollars.

Similar anticommunitarian logic characterizes corporate activity more generally. Corporations are legally required in some countries to be beholden only to shareholders, not to the places where firms are located. Consider how an Apple executive recently responded to criticism of the company's outsourcing of manufacturing jobs by replying, "We don't have an obligation to solve America's problems. Our only obligation is making the best product possible."[49] The implication is clear: Apple has no commitment to any community, not even the imagined community of the nation-state. With such a commitment it would be hard to justify the uprooting of factories when another municipality, state, or country can offer a more attractive combination of tax breaks or the relaxation of workers' rights and safety regulations.

Because thick economic community manifests in the commitments between owners and workers in the places where they are located, it is threatened by the economic changes that demand fragmented, mobile populations. One UNESCO study traced the decline of place-based communities in part to owners and managers ceasing to settle among and socialize with lower-level workers,[50] a situation exacerbated by increasing job insecurity and enforced mobility. As urban scholar Thad Williamson and his coauthors have contended, such "economic dislocation entails the wholesale destruction of civic networks."[51] Unstable social webs provide a weaker sense of belonging, and economic precarity and worry likely discourage active volunteering and community membership. Indeed, residential stability has been found to promote a sense of belonging and prosocial behavior. In contrast, residential mobility, having to frequently dissolve social ties and adjust to new circumstances, leads people to identify less with their social roles and more with their personal characteristics.[52] The more residential mobility is economically promoted or enforced, the more citizens become individualists.

Economic localism offers a more communitarian alternative to the standard economic paradigm. Localists aim to mitigate the effects of the "flexible economy" and of social fragmentation; they seek to reconstruct economic community as "formed in collective action based on place."[53] The intended effect of measures such as community-supported agriculture, local currencies, democratically governed businesses, locally sourced energy, and collectively held property through community land trusts is to sustain an alternative economy more rooted in relational interdependence and better responsive to local needs. The movement's goal is to help people meet more daily needs through cooperation, sharing, and more circumscribed competitive practices—farmers' markets, for example, have economic competition without harm to social relationships. Localism aims to diminish the degree to which the disparities of power, cutthroat competitiveness, and single-minded pursuit of profit characteristic of corporate capitalism define economic exchanges. Hence, economic localism lies in the middle between the more thickly communitarian practices of mutual aid and impersonal networks of economic exchange.

It is important to distinguish localist forms of economic sharing from the much hyped "sharing economy" composed of firms like Zipcar, Airbnb, and others. As Internet scholar Russell Belk has noted, exchanges on these networks do not really constitute sharing but rather are short-term leasing arrangements.[54] Such exchanges, as well as users' sharing of personal data and creations on websites like Facebook and YouTube, seldom contribute to the creation of a sense of solidarity or recognizable community. Rather, they might be more accurately viewed as commodity exchanges between network owners, consumers, and lay contributors. In the case of Web 2.0 participation, access to media or valued social space is traded for personal information. That is not to say such technologies could not better support more communitarian forms of sharing. Certainly they could, if they were owned by the communities using them and were less guided by the narrow pursuit of shareholder value or venture capital funding.

Thinner forms of economic community also exist. Businesses and consumers within a locality are bound into webs of economic interdependence regardless of whether they gather a sense of community from them. Similar interdependencies are strengthened between different geopolitical entities through trade agreements and forms of economic cooperation. For instance, the European Union trade agreement has more tightly coupled the economies of countries like France, Germany, and Greece and thus incentivizes the alignment of national economic policies, for better or worse. Similarly, college presidents and business leaders in the area surrounding Albany, New York, have organized an Economic Development Council with the aim of improving local economic growth. Although lacking the relational focus of economic localism, such efforts nevertheless aim to support the well-being of regions rather than merely individuals.

Political Community and Communal Justice

Thick political community exists to the extent that citizens self-govern via community institutions and to the degree to which disputes are settled in ways that preserve and stabilize relationships. The more impersonal and proceduralistic forms of politics that dominate most contemporary liberal democracies bind citizens into political communities in only the thinnest sense of the word: citizens are merely the constituents of representatives, and they are bound more by a shared constitution, legal framework, and some degree of civility than by strong relationships. Proposed communitarian or civic republican political alternatives, in contrast, place more emphasis on participatory self-governance.[55] Indeed, political theorist Robert Nisbet defined community as the product of people working together to solve problems, fulfill common objectives, and build the codes of authority under which they live.[56] Communitarians typically advocate for an increase in political community through the principle of subsidiarity: devolving political power to the lowest effective level. Thereby, everyday life is organized more through civil society institutions and responsive local governments than a distant, centralized state.

The thickness of the political community within institutions is related to their degree of impersonality. Thin political arrangements begin with the assumption that participants are self-interested strangers and even encourage them to act as if they are. Think of the competitive antagonism of court-directed divorce proceedings and inheritance battles or how the phrase "it's just business" is used to justify what would otherwise be relationally insensitive, if not unethical, behavior. Liberal political theory and therefore liberal political institutions, as philosopher Alasdair MacIntyre pointed out, are rooted in "the identification of individual interests ... prior to, and independent of, the construction of any moral or social bonds between [individuals]."[57] Communitarian institutions, on the other hand, govern behavior and resolve disputes in ways that preserve or enhance participants' relational bonds.

Nobel Prize–winning economist Elinor Ostrom has analyzed numerous cases of the latter.[58] From regulating a local logging forest in Japan to limiting catches in a Turkish fishery or providing irrigated water in Spain, numerous communities have successfully ensured that members share and sustainably utilize common resources for generations. Such efforts depend on members' capacity to communicate and negotiate conflict, on reliable mechanisms for developing trust through accountable monitoring, and on a sense of a common future. The Turkish fishermen regularly renegotiated rules and allowances, with monitoring made easy by assigning boats to neighboring time slots so that overfishing was immediately noticed by crews waiting to take a turn. Fellow fishermen understood the uncertainties and complexities of managing the fishery and knew the very real threat to their collective livelihoods if the resource were improperly exploited.[59]

Such institutions strongly contrast with the more bureaucratic systems common in liberal democracies, where setting quotas, monitoring catches, and policing compliance are handled by distant and impersonal organizations. The relationship between regulator and regulated, in the latter case, is more defined by competition or antagonism than longer-term cooperation. Impersonal bureaucracies, of course, are sometimes justified as less likely to be nepotistic or unfair. To anyone who has had extensive experience with them, however, especially around the globe, the view of bureaucracies as hedges against corruption is likely to seem suspect.

A more communitarian model of politics is likewise present in *bioregionalist* proposals for reforming affluent Western nations articulated by authors like Wendell Berry and Daniel Kemmis. Environmental philosopher Paul

Thompson has described the bioregionalist vision as premised on a belief in the importance of interactions with nature, food work, the wide diffusion of property ownership, and local interdependence for cultivating the habits and virtues that make good citizens.[60] To this end, bioregionalists call for a greater decentralization and descaling of industry, agriculture, and governance in order to scaffold a stronger felt connection to the environment and a sense of sociomaterial interdependence. They argue that place-based economies help citizens recognize a common stake in each other's welfare and support a politics guided in part by an acknowledgment of a shared fate. Contrary to the prevailing view, community is not conceived as rooted in some preexisting, nonconflictual consensus of values. Rather, political consensus (or, more likely, some mixture of compromise and concession) stems from and reinforces other dimensions of communality. This does not mean, of course, that very real material and ideological conflicts would disappear but only that structural arrangements could encourage negotiation via mechanisms that more often preserve than fracture relationships.

This process is visible on a smaller scale in the way some more communitarian societies approach the issue of justice. Consider geographer Jared Diamond's description of the case of a young boy being accidently struck and killed by a car in Westernizing but still very traditional Papua New Guinea.[61] Involvement of the police was minimal, limited to protecting the driver and his passengers from vigilantism. The dispute was resolved through mediation. Because the driver was at the time shuttling workers for a local office, his business manager met and negotiated with the boy's family. Instead of a drawn-out court case, open conflict was avoided through the provision of food and money for the boy's funeral as well as a feast where expressions of loss, suffering, and remorse could be aired and a collective commitment to peace could be made. My point with this example is not that accidents resulting in death would never result in manslaughter charges in thick communitarian societies—they probably should in cases of gross negligence. Rather, the practice of justice would be far broader than it currently is in Western nations, using jail time and other forms of retribution more judiciously.

Indeed, contrary to myths of communal harmony, thick communitarian societies are not characterized by the lack or avoidance of conflict but by the way conflicts are handled. Justice is premised not exclusively on rehabilitation, retribution, or deterrence, but on the preservation of relationships. Compare this with the Western system of criminal and civil justice, wherein relational ties are considered immaterial, do not exist, or are effectively severed in the process. Relationally restorative approaches to justice are motivated by the recognition that community members must continue to live peaceably with one another postconflict. Determining absolute winners and losers through an impersonal and antagonistic marshalling of evidence and argument tends to breed further resentment and distrust, undermining community stability. That is not to say that people living in societies that practice thick communitarian forms of justice do not hold grudges, have feuds, or otherwise act nasty to each other, only that their institutions aim to more constructively mediate interpersonal and civil conflicts.

Communal justice, moreover, rarely frames a dispute as existing solely between two individuals but involves those who are directly and indirectly affected. In Diamond's account, the compensation for the accident was handled by the driver's company. In different circumstances, it would have been paid by family members, to whom the driver would then owe a debt or nonmonetary responsibility. Thus, in cases of disputes and wrongs, community ties of mutual obligation are reinforced. In America, in contrast, the driver would have probably been fired and help from family would have been far from guaranteed.

Also essential to communitarian politics and communal justice are particular skills and dispositions helpful to resolving conflict. Among the most important is what political philosopher Benjamin Barber has called political talk, which "involves listening as well as speaking, feeling as well as thinking, and acting as well as reflecting."[62] Talking, for Barber, is not just a vehicle for expressing political interests but for bonding, understanding the positions of others, and establishing new shared meanings and imaginations as well. Clearly, the understandably tense ceremony involving the New Guinean driver and the dead boy's family would have been less effective absent the effort to listen, comprehend, and reestablish bonds. Likewise, larger political conflicts are likely to fracture a community if opponents see each other not as peers to be understood but only enemies to be vanquished. Such talk-based politics were visible in the Berger Inquiry, in which thirty-five "community hearings" held throughout Canada led to the relocation of a proposed oil and gas pipeline.[63] Attentive listening, however, is not well distributed throughout Western practices of politics, with some politicians and citizenries putting it into practice more reliably than others. The politically gridlocked U.S. Congress is one particularly negative example.

Without practice, the skills needed for negotiating conflict will likely atrophy, foreclosing possibilities for communal relationships. The first time most middle-class North Americans share a room with another person is

in their twenties or thirties. Lacking prior experience with sharing pos-
sessions or collective spaces, it is not at all surprising that many of them
view domestic disputes among roommates and partners as intractable, life-
draining burdens.[64] Elsewhere in the world, in contrast, room sharing is the
norm. Despite the inconveniences of such arrangements, people growing
up with them are likely to be better practiced at sharing, cooperation, and
tolerance for the presence of others. For instance, the anthropologist Jean
Briggs has described how an Inuit father met his needs for privacy by simply facing the back of the ice house.[65] As I discuss later, it is hard to imagine
how people could develop into capable political communitarians without
being forced by their domestic environment to learn such skills.

Community as a Shared Moral Order

Finally, community further involves the creation, maintenance, and evolution of shared moral values concerning the "right" way to live and how to
make sense of everyday practices. Those shared values and understandings
in turn influence people's self-understandings. As philosopher Michael
Sandel has argued, community members not only "profess communitarian sentiments and pursue communitarian aims, but rather ... conceive
their identity—the subject and not just the object of their feelings and
aspirations—as defined to some extent by the community of which they
are a part."[67] For relatively thin cases, such as in urban subcultures and
many online communities, identity formation more substantively *precedes*
relational communality: those with similar beliefs, hobbies, values, and
aspirations (e.g., middle-class hipsters) colocate so that others may *affirm*
already formed or desired identities. In its strongest or thickest sense, as
Sandel has contended, moral community is not chosen so much as discovered; it composes the ideological milieu built up by already existing
relations and practices from which individuals are constructed. That is,
identity emerges more *as a result of* already existing relationships and
social practices rooted in material, psychological, and political interdependence. This shared moral order is not necessarily totalizing, uncontestable,
or unchanging but, nevertheless, strongly influences the moral decision
making of members.

Thick moral communities also entail the shared norms, values, understandings, and identities that arise in groups where a "common good
beyond the sum of individuals' private interests is pursued and in which
individuals define their own interests and values in reference to collective
goods."[68] Members make sense of everyday practices in terms of shared

Box 3.2
Self-Management of Conflict and Community in Bethel

Established in 1985 in Hokkaido, Japan, Bethel House is a group home for people with schizophrenia and other psychological disorders. Because it is run according to a belief that the mentally ill become well through the process of reintegrating into a community as social beings and with a healthy skepticism of the paternalism typical in other psychiatric institutions, residents get to be comanagers of the home and its associated businesses. A local psychiatrist, Dr. Kawamura, works with residents to reduce their medication levels down to a practical minimum, a point where they can function socially and not be excessively bothered by voices or hallucinations. Especially when compared with the extreme isolation of most Japanese mental institutions, residents of Bethel find it to be a warm and accepting place. Tsutomu Shimono speaks fondly of being "treated as normal" and being able to live a "regular life" again. The home's youngest resident describes her housemates as "warm souls" who provide her hugs when she is upset, a practice of caring that she believes ought to be commonplace in her society but, lamentably, is not.

Bethel, however, is not and has never been an idyllic place. In Karen Nakamura's documentary depiction of the home, daily life is punctuated by both minor crises and conflicts. For example, a house member named Yamane created chaos in the community in the middle of a schizophrenic episode, insisting that he needed to go to a nearby national park in order to board a UFO and save the world. In another notable moment of conflict, a resident criticizes another named Yoshino for her body odor until she breaks into tears.

Nevertheless, Bethel's residents exemplify aspects of thick community in how they deal with such crises and conflicts. Yamane's episode was dealt with by a group meeting. Other members eventually persuaded him that he needed to see Dr. Kawamura to get a "UFO license" before departing, helping end the crisis without the involvement of the police. The incident with Yoshino provoked a group discussion about how residents could better talk to and help each other with their problems. Although far from perfect, the way in which residents of Bethel are able to put moral commitments to collective self-determination and communal care into practice, despite the challenges presented by mental illness, is inspirational.[66]

values and understand their own welfare as achieved through community life. The existence of moral community plays a strong role in supporting other dimensions of communality. Indeed, one sociological study of fifty urban communes of a variety of political stripes found that the strength of shared values and understandings of social roles was the most significant factor in determining members' level of commitment, feelings of belonging, and belief that their compatriots cared about them.[69]

Much of contemporary liberal thought, however, understands shared moral orders as the opposite of freedom, as a form of coercion enacted by community. As noted by Barber, liberty is typically understood as the absence of external constraint on individual choice.[70] Individual freedom is seen as accomplished through the development of sufficiently impersonal markets and bureaucracies as well as the procedural protection of individual rights. Achieving this moral vision, however, tends to be inimical to thick community. Indeed, some have argued that its logical conclusion is the destruction of any intermediate association or moral allegiance between the level of individual citizens and a centralized state.[71] That is, liberal freedom is ensured by bureaucratic and "invisible hand" institutions making sure citizens are kept sufficiently apart. For many readers, such a moral vision probably looks commonsensical, if not natural.

The difference between thick moral community and liberal individualism, however, is not really that the former impresses a shared moral order on citizens without their consent and the latter does not. Rather, they are simply different shared moral orders. Even highly individualistic forms of politics depend on the collective enforcement of liberal personhood by rendering unthinkable certain beliefs and activities. Liberal moral orders are visible in controversies over certain practices within ethnic, cultural, or religious communities: circumcision and the refusal of some deaf parents to implant cochlear implants into their children, among others. In these situations it becomes clear that a liberal moral order is not neutral but holds sacred certain avenues of exercising individual choice.[72] Liberal parents are not to deprive children of the right to choose for themselves particular aspects of their developing identity—whether sexual, sensorial, or otherwise—for the sake of group belonging. It is through these injunctions and others that liberalism exposes itself as a form of moral community.

This moral nonneutrality of liberal individualism is typically overlooked. The decision not to put a cochlear implant in a deaf child, for example, is actually not much different than the choice by one set of Toronto parents not to assign a gender to their child.[73] In both cases, a certain moral vision is being enacted on and via the child by their parents. The deaf parents value their child's belonging to the deaf community over his or her ability to hear. In the latter case, the child is allowed more space to wrestle with gender identity inside a popular culture defined by fairly rigid binaries. Yet did the child have a choice concerning whether to be forced to "discover" their own gender, much less about having to take a part in their parents' political statement? My point here is not to moralistically judge one case to be superior to the other but merely to point out that the deprivation of individual choice concerning shared values and practices is inescapable. Some moral order is already and always, albeit imperfectly, imposed on citizens by the mere fact of being raised by someone and within a culture.

Indeed, liberal personhood is itself not a choice but an enculturated reality. Consider how many middle-class Western parents aim to avoid tantrums by always presenting children with a choice: for example, "Do you want cereal X for breakfast or cereal Y?" Though this practice may often be effective, it nevertheless helps instill the particularly Western idea that individual choice is central to one's being. Indeed, psychological studies find that North Americans experience uniquely high levels of anxiety when unable to exercise choice over even relatively inconsequential matters.[74] The ubiquity of advertising and consumer culture further entrenches the idea that even the highly restricted individual choices offered via shopping represent the surest path to personal empowerment. As Gabriela Coleman has written, "today to liberate and express the 'authentic,' 'expressive' self is usually [interpreted as] synonymous with a life-long engagement with consumption."[75]

Liberal individualism is further naturalized via the narratives that dominate contemporary culture in technological societies. As philosopher Charles Taylor has argued, the unquestionability of individualism is enforced by "subtraction stories," which depict contemporary individualism as the ideal realization of humanity achieved by the subtraction of the "unnatural" influences of community and traditional forms of social organization.[76] At their most extreme, such stories give the impression that hiding within every member of a rural community, cohesive urban village, or hunter-gatherer society is a cosmopolitan (sub)urbanite aching to be free.

Consider the movie *Pleasantville*, a satire of televised depictions of 1950s family and community life.[77] The town is presented in black and white during the beginning of the film; characters only come into full color when they slough off some aspect of their conformity to Pleasantville's prevailing norms in order to better explore their individuality through various aesthetic, emotional, and sexual experiences. There is little wrong, from a communitarian standpoint, with the individual exploration of one's interests, personality, and sexuality. Yet, *Pleasantville* portrays the town's communal society as solely a barrier to individual self-expression, locked in a state of existential arrested development, especially later in the film when town fathers become increasingly oppressive in attempting to oppose the advancing coloration of community members. The film is a morality tale with a clear prescription: individualism as the cure for the supposed emotional, aesthetic, and sexual prison of community life. Indeed, one character's transformation occurs when they eat a colored apple, an allusion to the biblical tree of knowledge. Individualism is, hence, presented as the path to enlightenment, revealing the way to personal fulfillment and exposing the evils of the communal past and present to those willing to taste it. Looking upon the presumptions about the character of the good life and good society undergirding *Pleasantville* and other liberal "subtraction stories," it becomes clear that even individualism is a kind of moral community.

A common presumption in movies such as *Pleasantville* and other individualistic subtraction stories is that thick communitarian moral orders are immediately less free than moral individualism. Is this necessarily true? Rather than speculate over freedom writ large, it is probably easier to ask, "What is freedom under moral order X?" The moral order of most liberal democracies tends to privilege the economic liberties of property owners and capitalists, which results in the majority of population having little choice but to labor in nondemocratic, if not authoritarian, workplaces. While citizens in individualistic societies do not see themselves as obligated to maintain strong relational commitments as do people in more communitarian societies, they are also "freer" to suffer from anomie and loneliness, given that they are seen as responsible for arranging for their own needs for belonging and intimacy. The moral orders in thick communities emphasize different sets of responsibilities and freedoms. That members of immigrant communities feel obligated to support one another materially allows them to flourish despite generally not receiving living wages from the "free" market. Indeed, the very functioning of the community depends on maintaining a moral order in which members see the "right way" to live as entailing such obligations. It is unclear whether communal responsibilities are automatically more onerous than the compulsions and coercions of job insecurity, enforced mobility, and debt characterizing individualized, market society life, though they may appear unpleasant when viewed from a liberal perspective.

To some moral liberals, the compulsions of thick community seem not simply unfree but inevitably authoritarian. Although religious cults loom large as obvious counterexamples, one study of urban communes nevertheless found that a communal sense of belonging, commitment, and support tend to be inversely related to the existence of an authoritarian hierarchy.[78] That a communal moral order is necessary to an authoritarian moral order is hardly certain. Of course the way certain moral expectations are enforced in thick communities might feel tyrannical. In traditional small-town communities or bands of hunter-gatherers, for example, ridicule and other forms of shaming are used to ensure that no one appears to rise too far above everyone else.[79] Although probably unpleasant to the recipient and clearly restricting freedom of social mobility, the purpose of such behaviors is to level the social hierarchy, not render it more unequal.

A related concern is that the way thickly communitarian moral orders understand identity leads to bigotry. The intolerant traditional community is an archetypal story in popular media and understandings of history, not without some grain of truth. Sociologist Stephen Brint has argued, however, that exclusionary practices are not inevitable features of any community but a product of its "normative environment."[80] Indeed, the notably tolerant and politically progressive Quakers are as much a thick moral community as the Ku Klux Klan. Any discrimination against gay or transgender teens in a rural community would seem to have as much to do with the cultural history that led to the development of an intolerant normative environment than the place's status as a rural community per se. For example, perceptions of rural intolerance have been found to be no longer a significant factor in Swedish homosexuals' tendency to move to big cities. Moreover, social anthropologist Hans W. Kristiansen found a surprisingly high level of tolerance and acceptance of same-sex relationships in early twentieth-century rural Norway; the common belief that rural and working-class communities have been *universally intolerant* of nonheterosexual relationships, whatever its potential usefulness as an origin story for today's liberatory political movements, is simply not true. Indeed, gender and sexuality scholar Nadine Hubbs has maintained that the story of rural and working-class bigotry is told partly in order to reassure middle- to upper-class, urban whites of their own cultural superiority.[81] At any rate, such data undermine the view that the moral orders of traditional communities are necessarily intolerant of difference.

But does not the maintenance of a communitarian moral order demand the homogenization of identity? Certainly many thick communities are fairly homogenous, but it does not necessarily follow that they must be so.

Moreover, it is unclear to what extent homogenization is actually feasible, especially in less affluent areas. As sociologist Richard Sennett argued, the racial, socioeconomic, or physical segregation necessary to eliminate diversity requires some material abundance to accomplish.[82] In his view, the idea of the "purified community" is more a myth that powerful, affluent groups try to enact rather than a fact of community life—consider what it costs well-off whites to segregate themselves from everyone else. Community, Sennett contended, need not be premised on a moral order that privileges sameness but instead can be rooted in a love of distinctiveness. Consider how male-bodied transgendered people (*winktes*) were a tolerated and even spiritually revered group among many bands of the Lakota, in spite of otherwise rigid gender roles.[83] Another case in point is how the small rural town of Silverton, Oregon, elected the United States' first transgender mayor: Stu Rasmussen. The depth of residents' loving acceptance of Rasmussen's differences was demonstrated by an outpouring of support in the face of protests by the Westboro Baptist Church.[84] How difference is understood within a communitarian moral order could be not unlike C. S. Lewis's description of an ideal church: "[It] is not a human society of people united by their natural affinities but [one] in which all members, however different ... must share the common life, complementing and helping and receiving one another precisely by their differences."[85]

Dichotomous thinking regarding community and conformity overlooks how intolerance and policing of deviance have evolved rather than disappeared throughout recent history. Nationalism is an outgrowth of the growing importance of the nation in the face of declines of local forms of belonging. The egalitarian myth behind many conceptions of national identity frequently hides the real socioeconomic inequalities that pervade most countries. All the rhetoric regarding individuality within liberal, market-oriented moral orders obscures how advertisers enforce consumer conformity every day; being made to feel deviant for being overweight or out of style provides the impetus for consumerism. Finally, consider how the American justice system retains moral legitimacy among many citizens, despite disproportionately targeting African Americans and other minorities. The dominance of a liberal moral order in places like the United States has not meant that conformity is no longer enforced or deviance policed.

Finally, there is no clear reason why enforcing conformity with respect to identity or morally right behavior is prima facie a bad thing. Few would oppose informal community mechanisms for policing deviance when directed at antisocial behavior, uncontrolled expressions of anger and aggression, or other activities that can harm community members. Even proposed alternatives to thick community that take the ostensible anonymity and social and aesthetic inexhaustibility of metropolitan urban life as a model would require eliminating or policing anyone whose practices or ways of life were in any way incommensurable.[86] Such a society would only be possible by implementing norms, institutions, and sociotechnical systems to ensure that communities do not become *too thick*, and hence, tolerance would inevitably be extended to only certain kinds of differences.

It is far from certain that thick communitarian moral orders are necessarily less free, more authoritarian, or intolerant of difference. The egalitarianism of a form of social organization has more to do with its thickness as a political community and the presence of social justice–oriented moral values rather than the existence or nonexistence of a strong moral order. Indeed, proposals for more local and participatory models of politics, like Barber's "strong democracy," are as committed to increasing democracy as encouraging people to act and see themselves more as communal citizens than unencumbered selves.[87] At the same time, I have shown how even liberal individualism involves freedom-limiting moral injunctions and processes of enculturation. The choice for members of technological societies, therefore, is not between oppressive thick community and unequivocally free individualism, but rather between myriad combinations of communal and individualistic freedoms and moral responsibilities, ensured by institutions running the gamut from authoritarian to democratic and rooted in understandings of "we" ranging from open and difference loving to insular and exclusionary.

The take-home message of this chapter is that any manifestation of community, from online forums to a tribal village, can be evaluated as thick or thin in terms of seven dimensions of togetherness as well as its thickness across multiple dimensions (see figure 3.1). Thicker communities would incorporate dense and interwoven social ties; strong norms of reciprocity and mutual aid; opportunities for talk; the psychological or symbolic sense of community; economic interdependence; functioning systems for self-governance and conflict resolution; and a strong shared moral order.

This multidimensional view provides a more straightforward basis from which to recognize the thinness of forms of community like networked individualism, side-stepping stale arguments about which manifestations of togetherness are "authentic." Networked individualism is premised on diffuse and fragmented social ties, specific reciprocity, and a sense of belonging to a fuzzy rather than a coherent social entity. Even though

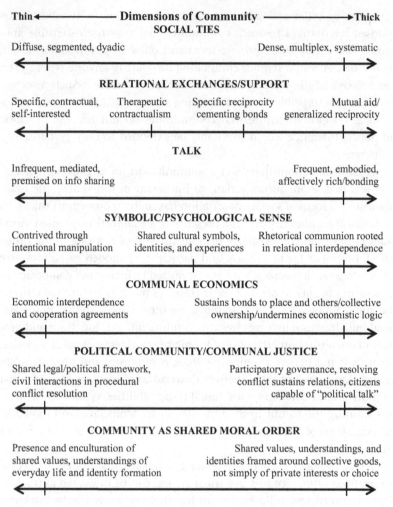

Figure 3.1
Dimensions of Community

networks provide some degree of genuine communality to some people some of the time, they do not score very high in their economic, political, and moral dimensions. Symbolic communities such as the nation or categorical identity groups are very thin when considering the density, multiplexity, and systematicity of social ties, being thicker in terms of shared cultural patterns and experiences that can result in a psychological sense of social connection. That some people find themselves unsatisfied by the provision of belonging in technological societies is likely due to them being insufficiently engaged in a sizable fraction of the relational, symbolic, economic, political, and moral dimensions of social life.

This multidimensional conceptualization, moreover, helps maneuver around the neverending debate over whether community, in the abstract, is inevitably intolerant. Political theorist Iris M. Young, for instance, has rejected community as an ideal on the grounds that the tendency to see it in terms of identity results in differences being labeled as deviant or pathological.[88] She is no doubt correct that the equation of community with shared ethnic, racial, national, or sexual identity has often led to unjust forms of exclusion and oppression. However, on-the-ground practices are more diverse than is captured by idealized, abstract visions of community. Indeed, feminist scholar Judith Garber has argued that although "community [as an abstraction] is deeply problematic; in practice, it may actually serve women more often than we think."[89] Indeed, it would be the height of academic condescension to write off the efforts of citizens to establish more communally responsive forms of policing or poor minorities banding together to form a worker cooperative because community in the abstract appears uncomfortably compatible with identity-based chauvinism.

To be fair, critiques of community as an ideal do have merit in sensitizing people to the risks of belonging in the absence of broader political community. As DeFilippis and North have argued, community's emancipatory potential lies in a politics that embraces rather than denies conflict.[90] It is the effort to deny conflict that leads citizens to attempt to unmoor community from any recognition of human diversity or plurality, a denial that, according to political philosopher Hannah Arendt, would undermine political action and the formation of political community.[91] It is the allure of togetherness-as-tranquility or belonging-as-shared-identity that drives the exclusion of those who are different. One witnesses such notions in the self-segregation of affluent whites to bucolic gated neighborhoods and various forms of ethnicity-based nationalism.

Although identity-based idealizations of community are unlikely to completely disappear anytime soon, they could be tempered by what David Brain has called the *ideal of civility*.[92] That is, one can value subgroup affinities in the context of broader public ties characterized by trust, tolerance, and mutual respect. In other words, it is a mistake to assume civility and community must be independent or opposed to one another. A shared valuation of the ideal of civility can serve as a basis for moral community as much as ethnicity, nationality, or any other collective identifier. Moreover, democratic ideals and practices, at their best, provide a rationale and means for constructively combining group interests with collective purposes.[93] To

best avoid the risk of intolerance, advocates for thick communitarian technological societies would be wise to craft proposals and actions so as to
avoid weakening the practice of democratic civil society.

In the following chapters I apply the multidimensional concept of
community developed here to the analysis of technologies that affect the
chances for rebuilding some of what has been lost, protecting what is now
working, and imagining new possibilities. How do artifacts, large-scale systems and infrastructures afford or constrain the enactment of thick or thin
practices of community life? Does a technology promote social connection
through diffuse networks or dense webs? Does it undermine or support
the recurring practices of material and social exchange that cement social
bonds? Does it enable citizens to escape personal and political conflicts or
encourage them to engage and negotiate? Does a technology contribute to
the building of an individualistic moral order or a communal one? Through
such inquiries, I develop an outline of what communitarian technologies
accomplish and what more communal technological societies could look
like.

4   Suburban Networks and Urban Villages: Community and the Built Environment

Technologies have political implications, differing with regard to which practices of community they support. They affect who gets what kind of social belonging. By the end of the following three chapters, it will be difficult not to see much of technological civilization as effectively legislating practices such as networked individualism on its members. Here I put my multidimensional conceptualization of community to work in analyzing a range of technologies: from city streets and heating stoves to Little League and techniques for sleep training infants. I examine each technology with regard to the dimensions of communality it enables or restricts, thickens or thins, however subtle or partial in its effects.

It may appear strange at first to call such a wide range of things technologies. Most people tend to equate technology with handheld gadgets, and often only recently developed ones at that. Technology encompasses more than just artifacts, however, but techniques, systems, and organizational structures as well. In fact, gadgets themselves generally rely on these other technological forms to work. Smartphones cannot function without communication, retail, and industrial networks. They have little utility, moreover, if users cannot perform techniques like swiping and texting. Finally, citizens would be less likely to own a smartphone absent the overall structuring to everyday life that gives them reason to want or need one, including workplace culture. Similarly, the cry-it-out method for sleep training infants is a psychosocial technique that relies on its own set of supportive artifacts, infrastructures, and organizations: from separate bedrooms with self-standing cribs to systems for distributing strategies and information through pediatric networks. Finally, something as large as a city can likewise be considered a technology: its infrastructures order everyday life and residents' access to goods and services. Contemporary cities tend to demand the use of certain artifacts, namely automobiles, as well as organizational techniques, including single-use zoning and strategies for avoiding contact with strangers.

My analysis is in the same mode as other philosophies of technology.[1] Centering such work is the observation that technologies are neither wholly determinative of human activity nor totally value-neutral tools. They are inevitably imbued with social and political values by their designers and individually influence people's patterns of thought and action in recognizable ways. The strength of a technology's influence varies, as Peter-Paul Verbeek has noted. It may "legislate" or "force" certain behaviors, a phenomenon illustrated by park benches that are built to make lying down impossible, preventing homeless people from sleeping on them. Others "persuade" us to act in certain ways, such as when the absence of sidewalks next to a busy road discourages walking or when the comfort of an air-conditioned den results in less motivation to leave the house on a hot summer day. Technologies can also "seduce," as do companion robots' feigned emotions that tug at users' heart strings.[2]

More broadly, the additive and multiplicative effects of constellations of technologies influence users' beliefs, habits, and expectations. They are "forms of life."[3] People who achieve belonging via social media and meet their material needs at Wal-Mart are not merely users and shoppers but beings whose very understandings of the world are shaped by such technological experiences. The more daily living entails negotiating interactions and exchanges through distant global-scale networks, the more such practices become taken for granted as normal. Every technologically mediated experience is a data point from which people extrapolate theories about the nature of reality. Dissimilar technologies support different meaning-giving "patterns" of everyday life.[4]

Not every user, of course, extrapolates the same sets of meanings, beliefs, and understandings. There is usually room for what technology scholars call "interpretive flexibility."[5] Nevertheless, the biases built into various technologies and sociotechnical systems are likely to have a significant influence on most people, whenever counteracting structures, practices, and ways of thinking are either absent or insufficient.

Where I depart from previous philosophy of technology is by explicitly focusing on community. Philosophers like Langdon Winner and Albert Borgmann have mainly emphasized the role of technology in democratic politics and in supporting the "focal practices" that give people meaning. To that end, Borgmann has depicted local baseball diamonds and farmers' markets as desirable alternatives to contemporary commodified living via sports video games and supermarkets.[6] Richard Sclove has focused primarily on the political dimension of community, defining communitarian technologies as those that "help to establish or maintain egalitarian, convivial, or legitimately hierarchical social relationships."[7] Although I will cover some of the same ground, my analysis is broader, integrating many dimensions of communality.

In the following chapters, I provide a systematic illustration of how broader cultural shifts away from thick forms of local community toward networked individualism have been sociotechnically supported. However, I aim to avoid overemphasizing the role of the usual suspects: digital devices and the automobile. I will pay considerable attention to the more subtle thinning of community accomplished via parenting techniques, networked energy and retail infrastructures, overly centralized organizations, and a range of other tools and gadgets. The turn toward online connection via smartphones and social media is really more a symptom and exacerbating factor of the already poor support available in the twentieth century rather than a primary cause. Indeed, the amenability of urban environments to thick community changed considerably during that time frame. As such, urbanity is a good place to begin my analysis.

***

Urban technologies have always been perceived as important shapers of community life. Historian Howard Gillette has well described how social reformers have frequently attempted to realize policy goals through urban redesign. Champions of the early twentieth-century City Beautiful movement hoped that grand civic monuments would improve social harmony, and mid to late-twentieth century experimenters believed low-income housing projects would defuse the pathologies of poverty and make the poor more like their middle-class brethren.[8] The history of attempts to build civility and community into urban form, however, is littered with unrealized aspirations and unintended consequences. Plans for communitarian "garden cities" morphed into suburbia, and efforts to provide neighborhood centers through commercial shopping spaces evolved into today's regional malls.

Even though urban form recognizably coshapes local social activity, the sheer complexity of everyday life makes it difficult to discern the causal connections. Built environments are large-scale social technologies that influence the character and dynamics of the flows, rhythms, and practices of daily life. How is networked individualism reflected in and supported by city spaces and institutions? Which urban technologies seem to better afford thick communitarian practices? Specifically, I ask: How do different arrangements of homes, buildings, and streets vary in the degree to which they support face-to-face public sociality, frequent interactions among residents, and opportunities to cultivate symbolic and rhetorical community? How do built environments influence the growth of social capital? Finally, what role do institutions of shared governance and conflict resolution play in the development of urban community?

Urban Form, the Geography of Ties, and Public Sociability

The most significant influence of urban form on communality is its effect on the first dimension of community: the geometry of social connection. The distances set by urban infrastructures between different spheres of everyday life enact a subtle but powerful force on residential relations. Typical suburban development, as sociologist Ray Oldenburg argued, appears as if it were designed for networked individualism.[9] Not only are nodes, whether homes or businesses, set at very low densities, but different aspects of daily living are functionally and spatially segregated. Large travel distances separate home ties from those maintained at work, school, or elsewhere. Suburban environments are large, "coarse-grained" networks.[10] Residents shuttle between large blocks of detached homes, shopping and strip malls, and business parks via high-speed freeways and highways, not unlike packets of data moving between distant server nodes. Residential areas become places for isolated domestic repose, not thick community. As M. P. Baumgartner has noted, suburban ties tend to "encompass only a few strands of people's lives."[11] Diffusivity, uniplexity, and the lack of systematicity are built into suburban form. Attempts to realize dense and multifaceted social webs are effectively prohibited by the limitations of time and space posed by sprawl and discouraged by the extent to which sociogeographic units below the scale of the city as a whole (e.g., neighborhoods) matter little for everyday life.

Within residential suburbs, the barriers to density, and to a lesser extent multiplexity and systematicity, are sometimes mitigated by the existence of shared public spaces. For instance, the surprisingly vibrant social environment that sociologist Herbert Gans found in his study of the original Levittown suburb likely had something to do with the provision of a playground, school, and community pool in every neighborhood.[12] Given that most people moving into these neighborhoods at the time were young families, the existence of public places for family activities provided centers for social activity. Such centers are less and less common in contemporary suburbia, however, which has followed a trend toward increased diffusivity. Community pools have been shunned in favor of private backyard versions as costs have come down for the latter and the former were racially reintegrated, and the increased perception of crime or "stranger danger" means that youths more often play indoors.[13] And the aging demographic of suburban residents means that child-oriented amenities connect fewer neighbors with one another than they did during the baby-boom era.

The relationship between the communality of social ties and urban form is further complicated by other intervening cultural and economic factors. As discussed earlier, urban metropolises were widely seen as anti-communal in the late nineteenth and early twentieth centuries. Yet, this perception may have had more to do with the rapidity of industrialization than with urbanity per se. A tenfold increase in population likely contributed to the high levels of transiency in New York City between 1830 and 1875 described by historian Kenneth Scherzer. Although accounts of mutual aid during epidemics and evidence gleaned from wedding guest registries show that neighborhoods still factored into people's everyday social lives, many New Yorkers of the era were already networked individualists pursuing "aspatial community."[14] Regardless, the rapidity and scale of the massive sociotechnical changes occurring within cities in that era make them unreliable points of reference for theorizing about urban community writ large.

Contemporary life too often presents a similarly distorted picture of the potential for urban community. Late twentieth-century North American urbanism is the result of affluent whites fleeing racial integration and the destructive practices of "urban renewal" that were partly motivated by modernist architectural agendas. Decades of redevelopment and neglect have produced cities more like suburbs than traditional urban neighborhoods. Whether in the form of low-income projects, high-income condominiums, or corporate offices, downtown development often means erecting high rises surrounded by parking lots and interspersed with strip mall shopping centers. Such areas, thereby, can be as coarse-grained as suburbia. For instance, by 2000 some 22 percent of downtown Hartford was dedicated to parking lots, and the city had a population density not much more than the sprawling suburb of Levittown, Pennsylvania.[15] The results are as one would expect: automobility is enforced; low densities are ensured; different use areas are highly segregated; and the land between segregated areas becomes "dead space" to be passed through as quickly as possible. Few consider walking through parking lots to be a desirable leisure activity. This building style is increasingly the norm even outside North America, albeit sometimes without the parking lots. Indeed, massive gated compounds of

residential high rises is increasingly the norm in China. Sidewalks lined by
security walls rather than stores and cafés, and open spaces in the shadow
of high-rise blocks, although efficiently segregating functional areas, are
poorly amenable to the practices of leisurely walking and sociable loitering
that support communitarian urbanity.[16]

Traditional forms of urbanism seem more supportive of thick geome-
tries of social ties and communal talking than suburbia or modernist cities. Jane Jacobs described how the medium-level density, mixture of uses,
small blocks, and architecture of buildings in Greenwich Village enabled
the "ballet of the sidewalk."[17] Such features focused residents' attention on
the street and put them in frequent interaction with one another. Neighbors were not necessarily intimately connected, but they were embedded
in dense and multiplex webs of public acquaintance. As Gans described
Boston's West End low-income urban village, "Everyone might not know
everyone else; but, as they did know something about everyone, the net
effect was the same."[18] Merely recognizing and having some information
about their neighbors provided the grounds for communality: a sense of
belonging, lessened barriers to serendipitous social interaction, and so on.
A similar richness of urban community life along with comparably supportive design features have also been found in the resident-constructed,
mixed-use settlements, or *favelas*, around Rio, albeit mostly prior to more
recent takeovers by drug cartels, increased subjection to state violence and
demolition, and gradual conversion into tourist destinations.[19]

Many architectural studies lend support to Jacobs's arguments about
urban form and public sociability. Mixed-use blocks in Cambridge, Massachusetts, promote more vibrant street life via a variety of local independent
businesses coupling publicly visible interiors with ample exterior seating on
wide, shaded sidewalks; William Whyte similarly emphasized the importance of ample seating, trees, and sunlight, as well as proximity to well-
traveled streets, in creating vibrant squares that attract a semidisorderly
mixture of pedestrians, street vendors, and local eccentrics.[20] Such design
features are almost universally present in traditional European urban form,
namely in a multitude of small and medium plazas located at the intersection of major pedestrian routes, surrounded by residences, shops, eateries,
major cultural landmarks, and political institutions. One particularly notable enablement of café culture is visible in Barcelona: buildings at intersections in some neighborhoods have beveled off, or "chamfered," corners,
providing ample room for people watching and social interaction exactly
where pedestrian traffic already tends to cluster. City residents, however,
do not take full advantage of this design feature: many of these chamfered

corners in Barcelona have been turned into parking lots—a move some city leaders and residents must have considered progress.

One, however, need not live in a large city to experience communal urban form. Anyone who has visited small towns in Europe has had the experience of standing in what appears to be a medium-density mixed-use city, only to walk a couple blocks and find forests and farmland: the urban–rural boundary is sharply demarcated. Rural living need not imply being isolated on a farmstead or acreage. Rural development could just as easily be inspired by urbanist thinking as suburban logics, creating pockets of density and largely preserving open spaces.

Regardless, forms of public sociability are generally absent within suburbia and sprawling rural acreages. High rates of depression among suburban housewives arose partly because the "low population density and the loss of natural daily social gatherings on the porch, the street, or the corner drug store made sharing experiences and ventilating problems more difficult."[21] In such places, television soap operas, the telephone, and social media are sometimes sought out to make up for lost opportunities for public gossip and exchanges of social support. Social networks no doubt do form in suburbia to a degree, but the distances between homes and the typical lack of sidewalks, front porches, and quality common spaces are strong material persuasions against frequent outdoor neighborly interactions. Suburban apartment complexes, which have the density to potentially support thick social networks, tend to be random clusters of large buildings interspersed with parking lots. Given that most residents fail to find an evening walk around a local parking lot to be all that interesting, few are likely to linger outside to chat with neighbors. The lack of mixed uses, or much else of interest to see, means that residents are typically out on suburban streets within the relative isolation of an automobile, preventing growth in the kinds of public social activity that might seduce residents away from their dens and back patios.

Developers of planned communities and suburban developments typically attempt to substitute a symbolic sense of community in lieu of the structural and practical arrangements that traditionally held communities together. Herbert Gans noted that the aesthetically homogeneous design of Levittown was an "architectural means for creating unity and cohesion that did not exist in the social system."[22] Suburban developments are designed and advertised to evoke the feeling that one is entering a bucolic semirural community. Widely derided as ersatz or counterfeit by critics of suburbia, this kind of symbolic community exists not as an emergent phenomenon rooted in social and political interaction or interdependence but as a marketing device: the branding of community as "Pheasant Valley" or the like.[23]

The enablement of communal interactions within high rises is often just as poor. One review of the architectural literature associated high-rise living with lower levels of prosocial behavior, with connections between residents being relatively rare—especially between floors.[24] Despite the hopes of modernist designers that apartment floors could come to functionally replace city streets, high-rise residents are more likely to seek out a sense of community through work or school connections. In the notorious St. Louis Pruitt-Igoe high-rise project, where all thirty-three buildings were razed in the mid 1970s, residents tended to eschew friendships within the building, preferring to retreat to their private apartment and cultivate more distant friendships elsewhere.[25] Such asociality no doubt contributed to internationally infamous levels of crime and general destructiveness within the development. At any rate, the sheer expense of floor area and very real physical constraints make it difficult to build high rises with large amounts of frequently traversed, lively, and highly visible semipublic spaces. There are few, if any, cases of a high-rise building functioning as a Jacobsian "neighborhood in the sky."

Features of medium-density urban spaces, in contrast, assist in the development of community at multiple scales. Margarethe Kusenbach distinguished four subneighborhood levels of communal interaction and attachment in her study of the Melrose and Spaulding Square areas of Los Angeles: microsettings, street blocks, walking-distance neighborhood, and the enclave.[26] Microsettings exist at the level of adjacent homes or apartments, facilitating friendly recognition and neighborly exchanges of favors. They are supported, but not guaranteed, by shared facilities (including laundry and mailboxes), communal areas, and the arrangement of space to encourage frequent passive contacts through similar routines. In addition to the sidewalk socializing discussed above, community at the level of the street block and walking-distance neighborhood is built up via informal collective events and social organizations. Get-togethers like block parties as well as the presence of neighborhood associations and "block captains" supplement sidewalk social life. The neighborhood enclave, finally, is as much symbolic as material, manifesting in collectively held identities that are spatially rooted in local landmarks, businesses, and public ritual events. For example, annual pride parades and fundraising events provided Melrose gays and lesbians the opportunity to rhetorically celebrate their local community subculture. Even this symbolic level of local community relies on compatible urban form: parades are typically infeasible on sprawling, low-density suburban roads or divided highways, and communal celebrations require visible, well-trafficked, and suitably large public places.

Design movements like new urbanism aim to recreate the kind of vibrant street life described by Jacobs through a revitalized version of traditional urbanism: buildings set close to sidewalks, large front porches, mixed-use zoning, higher-than-suburban densities, and ample public space. Studies have found that new urbanist design features do promote a sense of community and increases in local pedestrian-oriented socializing, though by how much and how reliably remains contested.[27] Moreover, many new urbanist efforts to artificially recreate the "ballet of the sidewalk" fall short: They fail to seduce a large chunk of residents out of their cars, and end up being slightly more photogenic, upper-middle-class suburban enclaves. There are a range of intervening variables between physical form and the existence of community, so alterations to the design of buildings and neighborhoods alone cannot be counted on to foster togetherness.[28] At best, they are supportive enablers or persuasions toward thicker communal social webs whose effectiveness, in turn, depends on a range of other socio-technical infrastructures.

Destinations, Economies, and Social Capital

Vibrant communal street life likely requires not only suitable densities, sidewalk design, and mixture of uses but also neighborhoods that contain desirable destinations. The ubiquity of "residential areas where there is so little of interest outside people's homes [means] that the privatization of life is no longer optional but spatially enforced."[29] Amenities like pubs, cafés, and local businesses form the roots of local economic community, not merely the social or symbolic dimensions of togetherness. Local economic community is undermined in suburban and similarly coarse-grained urban spaces because zoning requires that retail be clustered in distant malls and big-box stores, placing middle- to upper-class jobs in sprawling office parks; neighborhood-level economic activity is literally prohibited. Once residents are forced to drive or take the bus to get to the grocery store, it hardly matters to which location they go; all are fairly equally detached from their residential community. As a result, residents have little opportunity to develop multistranded relationships with store proprietors, fellow shoppers, or workers. Without this multiplexity, the relationship between the businesses and shoppers is likely to remain defined by narrowly acommunal, contractual logics of economic exchange.

Some planned residential developments include local shopping centers, but they often still fail to spur the development of economic community.[30] Large regional malls and outlets continue to persuade or seduce residents with promises of lower prices, more variety, and free parking; the typical resident already must commute to work outside the neighborhood, which provides a psychological incentive for them to shop elsewhere rather than park at home and walk to local stores; and retail staff are very often paid too low a wage to be able to live in such planned residential developments. Promoting local economies would require lessening the barriers to bringing work, home, and shopping closer together, not to mention targeting all the invisible subsidies going to malls and Internet retailers.

Urban environments supportive of local economies encourage both public sociality and the development of social capital through neighbors' exchanges of favors and mutual aid. Jane Jacobs described how residents in Greenwich Village would frequently entrust local store owners with holding their apartment keys for friends as well as how the proprietors contributed to collective safety via their "eyes on the street."[31] They were not merely businesspeople but "public characters" who would update parents on their children's actions, spread local news, and help resolve disputes. Local store owners have traditionally provided credit to the "unbankable," and local cafés and restaurants have served as places for receiving mail and hammering out business deals.[32] Other important local eccentrics are the political agitators who mediate between the neighborhood and higher levels of governance. Without public characters to stimulate communal sociality and social capital, everyday life is very different. Indeed, Jacobs recognized long ago that, lacking street-level community, people would have to rely increasingly on private social contacts for what such individuals provided.[33] Without vibrant forms of public sociality, people must turn more and more to networks for important exchanges and support.

Indeed, given the poor amenability of suburban spaces for public social life, residents foster local relationships and exchanges through other institutions. Voluntary associations, clubs, and other arrangements— especially in the United States—help realize the everyday sociability and social capital in suburbia that vibrant urban spaces accomplish through compatible urban form and economic community. Herbert Gans described Levittown as not a community "by any traditional criteria" but rather "an administrative-political unit plus an aggregate of community-wide associations." Suburban PTAs, Boy Scout troops, churches, and common interest or hobby groups do provide ways for neighbors to get to know one another and provide each other social support. As important and desirable as such groups are, by themselves they do not guarantee a cohesive neighborhood-level community. Voluntary associational life within Levittown, although perhaps enviable by today's standards, amounted to a "loose network of groups and institutions" rather than a coherent social entity.[34] That said, there are important differences between being on the PTA and attending a book club. Voluntary associations strongly tied to local problems and concerns likely contribute more to a sense of community than do discussions of literature.

Neighborhood-level businesses also develop social capital in their role as local centers for talk. They are *third places*: sites of significant social interaction outside the home and workplace.[35] Third places can be hair salons, grocery stores, shops, cafés, pubs, libraries, or community centers that serve as facilitative spaces for chance meetings, playful conversation and gossip, and inexpensive relaxation. At their best, they support social mixing and are somewhat leveling of distinction. Good examples are the Eagles and Veterans of Foreign Wars taverns omnipresent throughout small towns in Montana, where one often finds college instructors rubbing elbows with cattle ranchers, ditch diggers, and local professionals—including lawyers and pharmacists. In such welcoming spaces, residents know they will both be recognized and recognize others.[36] Similar to vibrantly social neighborhood sidewalks, local corner stores, cafés, and parks offer residents, including stay-at-home mothers, retirees, and the unemployed, a place where social connection is a public good: it is the organic byproduct of the flows and rhythms of everyday life rather than the responsibility of individuals entrepreneurially operating social networks. Daily life in neighborhoods without third places is more anonymous, which can be desirable for those wishing to escape the gaze of friends, family, and acquaintances but comes with the risks of isolation and impersonality.

Contemporary urban designs, however, stymie the development of third places, because single-use zoning prohibits them and physical barriers to walking undermine their economic viability. Given that patrons often buy little more than a cup of coffee, being in walking distance can be an important factor for a third place's viability. Most suburban residents would acknowledge that the cost of gasoline, time constraints, and the frustrations of traffic and locating parking persuade against frequent small outings, much less "just popping by."

In the absence of third place neighborhood pubs, nights out typically become more of a networked affair. Visiting "the local" is replaced by pub crawling among several bars or dance clubs. The pub crawl is akin to channel surfing, stopping by briefly to consume the atmosphere and moving on. Consider the way urban hipsters and other bohemians crash blue-collar bars to exoticize their culture like a nineteenth-century colonial anthropologist. In another sense, going out becomes a means of reaffirming one's personal community of networked ties. According to Robert Hollands "the ritualization of nights out has become, if you like, young adults' attempt to construct a modern equivalent of 'community,' or more correctly, communities."[37] Bars in this mode do not center community life within neighborhoods but instead compete to be a desirable node for fragmented networks of personal connections. This is often done through attempts to evoke patrons' sense of nostalgia by mimicking the aesthetics of the neighborhood pub, while actually lacking its sociological character. Regardless, most suburban developments do not offer a real choice between the neighborhood community of the local tavern and networked communalizing within downtown bars and nightclubs: zoning prohibits a neighborhood tavern from ever existing in the first place.[38]

Other Scales of Urban Community, Governance, and Conflict

It is important, however, not to overemphasize community as a neighborhood-level phenomenon. Thick communal ties, as noted in chapter 3, are systematic. Lower levels are incorporated into higher levels like Russian nesting dolls. Therefore, insular built environments, including gated communities and urban ghettos, stymie this facet of thick community. Although many planned developments lack the more obvious urban boundaries of gated communities (e.g., fences and security guards), property values and aesthetic homogeneity can be just as effective in isolating the neighborhood. The relative social isolation of the urban renewal neighborhood of Villa Victoria in Boston's South End, to take one example, was due not only to physical barriers (poor road connectivity with surrounding neighborhoods) but its architectural style as well, which distinguished it as a working-class barrio in a middle-class area. A similar kind of planning-enforced social isolation no doubt hindered the communal development of many twentieth-century low-income projects.[39]

The lack of systematicity, moreover, tends to lead to more unstable, pathological, and politically weak communities. Indeed, Gans found this to be the case in the Boston West End urban village he studied.[40] Although strong in street and peer-group community, residents lacked a sense of commitment to the larger neighborhood. Therefore, they were unable to mount much opposition to urban renewal processes that eventually destroyed their homes. Partly for this reason, urbanist Jane Jacobs advocated ensuring that urban locales were partly self-governing units incorporated into larger city districts and heavily criticized the mythos of the insular neighborhood. The boundaries of communities at any scale likely need to remain somewhat permeable to guard against becoming politically weak ghettos or suffering other isolation-induced pathologies.

Urban communities are made more systematic when the physical and symbolic boundaries of neighborhoods and districts coincide with the organization of public services and the reach of political institutions.[41] An urban environment becomes more psychologically meaningful for residents if it is not merely an incubator for social interaction but also an important space for obtaining needed services and doing politics. A case in point is community policing, which aims to reduce the degree to which officers and local residents feel alienated from one another—a growing problem as police forces become ever more centralized. In addition to potentially making police more accountable to citizens for acts of brutality or racial profiling, increased local oversight of policing renders neighborhoods more politically relevant to people's lives. Yet, most residential areas and buildings are not self-governing units or are controlled by homeowners' associations, condominium boards, or improvement districts. In the first case, residents are able to experience their locale, street, or building as a political community only informally. The latter institutions often only provide weakly democratic inklings of self-governance.

Homeowners' associations and condominium boards are usually neither integrated into larger scales of municipal government nor charged with a public responsibility; instead they are private nonprofit corporations dedicated to preserving property values.[42] Their institutional model is biased toward ensuring the stability of the exchange-value of homes over the use-value of the neighborhood, a priority visible in their enforcement of stringent systems of rules. Many associations ban political signs, satellite dishes, pets, and pickup trucks, and also enforce aesthetic homogeneity by setting limits on house color and acceptable renovations. One homeowners' association in Boca Raton, Florida, has even prohibited children from riding bikes on the development's roads or playing in common areas out of liability concerns.[43] Although residents vote on board members and rule changes, the design of associations renders them inevitably conservative; because they require that two-thirds or three-fourths of all homeowners agree for any given rule change, regulations put in place by the original developer are seldom altered. Given these democratic limitations, residents are less like citizens than shareholders. Unsurprisingly, they usually turn to litigation or selling their share (i.e., moving) when conflicts arise.

Quasipublic residential or business improvement districts are increasingly utilized technologies for creating small-scale political community in the United States, Canada, the United Kingdom, and South Africa. Within these districts special taxes or other mechanisms are used to fund services to supplement those offered by the municipality, including sidewalk repair, security, and street cleaning. Transportation scholar Donald Schoup has described how downtown businesses in East Pasadena, California, organized into an improvement district that began charging for on-street parking, plowing the subsequent revenues back into local improvements.[44] This arrangement enrolled residents and business owners into monitoring for violations, given that the revenues more directly benefited them, and provided financial resources for them to act on their local concerns. Most importantly, participants became engaged in the communal governance of a collective resource: parking. Improvement districts do not come without their problems, from exacerbating interneighborhood inequalities to being dominated by large landowners and assessment payers.[45] Nevertheless, they could be a starting point for more democratic, local-scale governance of collective resources.

A range of activities could be organized and governed in ways to better synergize with the symbolic or psychological community of neighborhoods. Sporting and recreational activities for children and adults would best support community systematicity if teams and leagues were organized so as to scale from the neighborhood level up to larger districts. Savings in capital and organizational costs, however, incentivize centralization. The Canlan Ice Sports Center in Scarborough, Ontario, for instance, organizes multiple divisions of beer-league hockey with dozens of teams in a massive facility housing four NHL-sized rinks. Although such a facility supports a thriving network of hockey enthusiasts at the scale of the city of Toronto, it undermines the possibility for citizens to experience a block, neighborhood, or district-level hockey community. Moreover, players must commute and are far less likely to participate in the governance of their recreational activities; they are relegated to the role of a consumer shopping for sporting opportunities. Municipal planning could better support community sport by ensuring the presence and proximity of inexpensive and locally managed basketball courts, outdoor hockey rinks, and soccer pitches along with pathways toward district-level competitions where participants would represent their neighborhoods.

Apart from the lack of opportunities to participate in and govern local institutions, urban form affects civic engagement more generally. The biggest disincentive against community engagement comes from the de facto requirement of commuting. Political scientist Robert Putnam found that "each additional ten minutes in daily commuting time cuts involvement in community affairs by 10 percent," and more recent studies largely confirm such findings.[46] Simply ensuring a more fine-grained distribution of workplaces and affordable housing would free up time for community as well as allow communal ties cultivated at work to more often support or reinforce residential community rather than compete with it. Moreover, economic homogeneity is highly correlated with political apathy and civic disengagement.[47] Separating affluent homeowners from renters and apartment dwellers obviates political conflict within neighborhoods by segregating socioeconomic winners from losers. The affluent are able to escape urban problems by running off to exclusive neighborhoods, and the poor are isolated from both the powerful and potential middle-class political allies. Enforcing a mix of varied housing types and incomes is an obvious counter to this process.

Urban form is not merely made up of infrastructures, zonings, and buildings but is tied into larger sociotechnical systems of construction, home ownership, and conflict resolution, which themselves affect the practice of community. Cooperatively owned and governed housing and community land trusts are promising alternatives to private ownership models.[48] Under these schemes, buildings or land is collectively owned through a community-governed corporation, which has the authority to intervene on the sale of housing units. This differs from condominium and homeowners' associations where sales are structured as traditional contractual exchanges between buyers and sellers. Narrowly contractual exchanges are less likely to evoke the sense that one is not simply purchasing real estate but also joining a community. Indeed, a longitudinal study of one of the first planned communities found civic apathy and free riding to be the norm.[49] Such behavior is probably not at all unexpected given that planned-neighborhood development frames community as an amenity to be purchased via an individual mortgage and association dues.

The atomistic character of buying into a suburban neighborhood, along with many of the other factors discussed thus far, helps drive what M. P. Baumgartner described as the suburban tendency toward a "minimalistic" moral order.[50] Suburban social relations tend to be defined by the avoidance of conflict and strangers as well as the absence of mutual aid. It is likely this morality of individualistic noninvolvement and nonconcern that led urbanist Lewis Mumford to deride suburbia as "a collective effort to lead a private life."[51] Other studies have found a similar moral aloofness in gated condominium communities and upper-middle-class market-rate housing cooperatives.[52] Residents rarely substantively interacted, and they resolved conflicts only indirectly by appealing to their homeowners' associations and representative co-op boards. Ironically, even the co-op, though ostensibly rooted in participatory principles, was dominated by a culture of avoidance and atomistic individualism. The moral order in such condos and co-ops seems hardly different, albeit more orderly and peaceful, than the one observed to dominate the low-income Pruitt-Igoe project.[53] Lacking the semipublic space around which neighborly interactions, mutual monitoring, and informal social networks could develop, all were defined by a culture of minimal social trust and, hence, a reliance on impersonal institutions like co-op boards or police to resolve interpersonal conflicts. Cooperative ownership, therefore, might be helpful but is ultimately insufficient by itself to overcome the barriers to community posed by atomizing urban form.

Collaborative housing, or "cohousing," models go beyond collective ownership by attempting to more directly foster community life within a building or set of buildings.[54] Cohouses typically have more semipublic spaces, like shared kitchens, terraces, gardens, and lounging space, that are collectively managed. Important design features include plentiful and highly visible public spaces, shared pathways to communal resources and sites of activity, scales and densities high enough to promote frequent interactions and mutual recognition as well as enough private and semiprivate space for residents to meet privacy needs, but not so much as to encourage withdrawal.[55] Beyond these material design elements, cohousing residents rotate responsibility for tasks such as cooking collective meals for their neighbors, cleaning, and routine maintenance. Cohousing supports the thick enactment of multiple dimensions of community at once: economic interdependence, political governance, talk, and the formation of dense social networks.

Another potential avenue for displacing economistic and morally minimalist logics with more communitarian ones would be to involve citizens and potential community members in the design and planning of their neighborhoods. A case in point is the infill development of Quartier Vauban in Freiburg, Germany, which was steered in part by a citizens group (Forum Vauban) in addition to Baugruppen, or building cooperatives.[56] Instead of being designed mainly by developers and planners guided by meager amounts of citizen "consultation" and their own (usually conservative) assessments of "what the market wants," citizens were highly engaged throughout the process—setting standards and developing their own plans. In the 60 percent of the neighborhood developed through building cooperatives, future residents worked directly with architects to design their future homes and cohousing units. Constructing Vauban was not merely the erection of urban form but itself an act of community building: engaging citizens with one another in conflict over a vision of a common future and over the codes and regulations by which they would be governed.

I would go so far as to argue that measures that spur local conflicts and disagreements about urban form as well as their constructive resolution help to thicken moral and political community. This may seem like a strange assertion, given that many people tend to equate community with a vision of tranquil harmony and togetherness. As Baumgartner has argued, however, conflict and care are often opposite sides of the same coin.[57] She found care and social support to be much more visible in the working-class neighborhoods where conflicts were also more openly confronted, even if not always constructively. Middle-class family members and neighbors were not strongly bound together by mutual aid and exchange and, at the same time, would sooner move out than air their grievances. Conflict avoidance parallels a lack of interdependence. Disputes over public space and actions, moreover, are probably only likely when residents are invested in or care about their community. Disagreement and contention, moreover, are great for forcing residents to express their care or, on the contrary, expose their apathy. Indeed, according to Gans, "Organizational formation and community activity generally are enhanced by conflict and crisis."[58] Chronic, poorly resolved conflicts, of course, can fragment communities or lead to social ills, but it is likely the resulting disengagement and apathy that destroys community rather than conflict itself. Opening up more local opportunities for disagreement and strongly democratic means of resolution is likely as necessary for thickening community life as compatible urban form.

Clearly, several connections tie together community, networked individualism, and urban technologies. More individualized forms of sociality are encouraged by the coarse-grained design of suburbia and post–urban renewal cities. Bereft of semipublic spaces and the vibrant street life supported by human-scale and mixed-use design, and lacking the prevalence and proximity of community-centering "third places," residents are more likely to look inward toward private domesticity and outward toward distant and diffuse social networks for interaction and support. These practices no doubt have their benefits, but without strong street- and neighborhood-level ties, mutual aid and informal means of conflict resolution have difficulty taking root. Hence, residents turn to impersonal and individuating

bureaucracies to perform the surveillance, social control, and conflict resolution that community ties and institutions traditionally provided. The
governance of buildings and neighborhoods is either enacted by distant
city governments or by weakly or undemocratic corporations, stifling
the growth of political community. Community as a symbol is too often
contrived through architectural homogeneity rather than through participation in political and social conflict, neighborhood-level recreation, or
rhetorical moments of communal celebration.

More communitarian urban spaces would be more than just redesigned
buildings, sidewalks, public spaces, and blocks. A sufficient density of desirable social destinations—such as shops, eateries, and taverns—would promote communal exchanges, thicker local economies, and multistranded
webs of social connection. The coincidence of the physical and symbolic
boundaries of neighborhoods with those of political institutions and service providers is likewise an important factor. Neighborhoods in larger
metropolises have more communitarian potential, especially psychologically and with respect to political community and the geometry of social
ties, when meaningfully incorporated into districts. Most importantly,
and in contrast to prevailing myths about community, conflicts within
communitarian places would be openly and constructively confronted
rather than avoided or delegated to more impersonal bureaucracies or
solved through legal action. Their moral orders and political practices
would be thicker than they are currently, requiring institutions that
encourage mediation and that develop rather than diminish social capital. Quartier Vauban residents, for instance, established a neighborhood
conflict-resolution group, which attempts to mediate conflicts toward
compromise and subvert the German tendency to always report problems
to the authorities.[59]

Urban technologies can only shape community life so much. In the
language of science and technology studies, urban spaces are flexible to
different interpretations. They are often appropriated in ways contrary to
designers' intents. One readily finds retired seniors and stay-at-home parents chatting on weekday mornings in the food courts of regional malls in
cities such as Edmonton, Alberta. They manage to turn such spaces into
quasi-third places, despite all the efforts by mall planners to make them as
unamenable as possible to loitering and relaxed conversation. Small-town
residents sometimes accomplish much the same in strip mall cafés and fast
food outlets.[60] Although many rural areas in America have been traditionally suburblike and diffuse in comparison to European villages, residents
often overcame spatial barriers in their commitment to visitation practices.

Essayist Wendell Berry described the southern version of such visitation practices as "sitting till bedtime," where neighbors would walk across their fields to sit, eat, and tell stories until late into the night.[61] Similarly, Baumgartner's working-class subjects sustained rich communal ties within the same suburb as their more anomic middle-class neighbors, a result also found by studies of the social environments of laundromats.[62]

However, the existence of such practices should not be used to minimize the very real barriers to community presented by certain urban technologies. For every senior or young mother meeting friends at a regional mall or fast-food joint, there are many for whom gasoline prices, time constraints, and traffic are persuasive disincentives. Contemporary rural residents are probably more likely to be "sitting until bedtime" in front of televisions or computers than with neighbors on their front porches. Just because some are able to realize thicker forms of communality in spite of real sociotechnical barriers does not mean everyone else can just pull themselves up by their social bootstraps and do the same. Not only is such a view naïve, in that it denies that barriers can be difficult for individuals to overcome, but it is inevitably a moral imposition, in that it sees the solution in adapting everyone to networked individualism. The existence of interpretive flexibility does not mean that the influence of technologies themselves is trivial but that it is mediated by the influences of other technologies, systems, and ingrained cultural practices. People must negotiate whole constellations of sociotechnical forces, persuasions, and seductions, which are often contradictory. Rather than closing down the question of community and technology, this observation simply alerts one to the near obvious: urban technologies alone, whether in built form or governance, are usually insufficient to make or break community. The roots of networked individualization run far broader and deeper.

Table 4.1

Urban Form's Intersection with Dimensions of Thick Community

Dimension of Community	Thinner/Barriers	Thicker/Enablers
Social ties	Coarse-grained suburban or urban renewal development; gated and walled communities	Mixed-use, semi-public space; Shared amenities that make daily routines visible
Exchange/ support	Voluntary associations (shared interest/lifestyle)	Coincidence of service boundaries; design supporting local business/public characters; voluntary associations (local needs)
Talk	Design for automobility rather than walkability: e.g., parking lots and other dead space	Third places; cohousing; mixed-use semipublic space with adequate seating; porches
Symbolic/ psychological	Aesthetic or architectural homogeneity	Coincidence of boundaries in recreation; supportive spaces for public rituals and celebrations
Economic	Regional malls and shopping centers	Mixed-use/local business; cooperative ownership; collective resource management through improvement districts
Political	Weakly democratic homeowners' associations, condo/co-op boards	Democratic business improvement districts, cohousing, building collectives; participatory planning; systems of conflict mediation
Moral	Community as purchasable amenity; myth of communal harmony in bucolic suburbia	Cohousing, collective ownership; emergent out of mutual interdependence and strong networks

5 Infrastructures and Organizations: Embedding Members or Networking Individuals?

This chapter extends the analysis of technologies' compatibilities with forms of community life to the infrastructures and other systems that organize the distribution of everyday goods and services. As with the built environment, infrastructural technologies vary in how much they arrange people as disembedded nodes on a network or as persons integrated within more coherent communal webs. Are users framed as genuine participants or mere atoms in an impersonal bureaucracy or market? Does the infrastructure support social interaction and longer-term relationships among those served by it? Does it incorporate several dimensions of communality, including collective governance, local ownership, and symbolic and psychological forms of belonging, among others? Similarly, does the organizational technology, at the very least, allow for goals and practices to be partly determined by those served by and composing those organizations?

Infrastructure

The introduction of indoor plumbing in the Spanish village of Ibeica altered the social fabric in unanticipated and ultimately undesirable ways. As political theorist Richard Sclove has noted, the story illustrates the unintended consequences of technological development, which could be more often avoided if innovations were more consciously and democratically governed.[1] More pertinent to the question concerning technology and community are the ways in which this infrastructural change affected the character of local sociality. Because the introduction of indoor plumbing made water available within residents' private homes, local women no longer frequented the public fountains and wash basins to do their laundry, socialize, and gossip. The men interacted less with the many children in the village who used to assist in the process of gathering water.

The result was an unweaving of communal bonds by the loss of community gathering places and practices that ensured high levels of social interaction.

This change should not be understood as simply a story of technological advance: systems of communal wells and indoor plumbing are equally technological. Where they differ is in how they configure social relationships. As philosophers of technology Adam Briggle and Carl Mitcham have put it, the infrastructural shift in Ibeica meant that residents were "simultaneously *disembedded* from the communal gathering of water and *networked* by a system of pipes."[2] Indoor plumbing makes individual homes the sites of distribution for water coming through large-scale water treatment facilities centrally controlled by municipalities or regional governments. The locus of distribution and control for a village well, in contrast, lies at the scale of local community—perhaps involving only a few hundred people. The wells are a site for not only small-scale social interaction but governance as well. Given a limited number of wells, villagers are forced to collectively devise some institutional arrangement for deciding who gets what water, when, and how. Furthermore, although the tools, materials, and techniques for building the village fountains were likely partly supplied via larger-scale sociotechnical networks, their use and interconnections with other networks manifest in ways that are more legible to residents. That is, they are transparent and more easily comprehensible as a visible part of the larger social web.[3] In contrast, indoor plumbing systems typically cannot be experienced at the neighborhood scale; relevant networks and nodes are usually too large to evoke a sense of interdependence between users. Water enters the home as an abstract commodity obtained by individual consumers in their purely economic relationship with the utility network. Where it comes from, where it goes, and how it gets from A to B are not things most water users are made aware of when opening the tap.

Not only did infrastructural change in Ibeica mean the loss of places that gathered and centered talk, exchange, local social relationships, and practices of self-governance, it also reinforced a thin conception of moral community, that of the unencumbered liberal self. This moral imaginary is reinforced by users' engagements with sociotechnical systems wherein the reality of utter structural dependence is overshadowed by an apparent experience of individual freedom and autonomy.[4] The experience of turning a tap and obtaining water whenever one wishes feels immediately empowering because of its convenience; one's ultimate dependence on large-scale sociotechnical systems is only clear upon reflection or when those systems break down. Such experiences are the building blocks of *technological liberalism*, the view that technologies unequivocally and unproblematically enhance human freedom via expanded opportunities for individual choice.[5] They feed the myth that individuals can and largely do "take care of themselves."

The myth of self-sufficient independence in the face of growing large-scale interdependence has only become more dominant over centuries of network building. Throughout American history, popular narratives have portrayed successive waves of development as accomplished through the enterprise of heroic individuals. Ironically, such individuals have always been dependent on vast sociotechnical networks: industrial systems producing the axes needed to clear forests, and railroads, canals, and irrigation infrastructures largely built or financed by governments.[6] The contradiction between individualistic myths and interdependent realities is sustainable in part because of the difficulties people face in comprehending a large, complex sociotechnical system as well as how they fit into its workings.

Most members of affluent Western nations have had indoor plumbing for some time, and communities have clearly persisted in spite of it. My argument is not that indoor plumbing and community are incompatible, but rather that thick community depends on at least some of the infrastructures of everyday life functioning more like village wells than indoor plumbing. Residents of Picton, Ontario, for instance, fought Canada Post when it proposed installing new mailboxes closer to their homes. They preferred frequenting the downtown sorting center where they could socialize with neighbors.[7] A variety of goods and services could be provided at the neighborhood or town scale, enabling distribution points to become focal points of social congregation.

The mere existence of collective, smaller-scale infrastructures, however, is probably insufficient in itself. Common laundry rooms in many apartment complexes typically fail to foster the development of communal relationships between residents already disposed to private living or committed to networked social connections. Nonetheless, there is communitarian potential in small-scale infrastructures done well. A resident-run laundromat founded in the Wentworth Gardens housing project in Chicago, for example, served as an important social space and a meeting place for community activists.[8] Although moving from an asocial, or even antisocial, laundry room to something like that enjoyed in Wentworth Gardens is no simple matter, clearly nothing of the sort could possibly exist if every resident is provided with en suite laundry.

As I outline below, an increasing number of infrastructures take the form of diffuse networks with high levels of centralization. These technologies individuate the experience of the infrastructural good or service: The user is treated as an atom or node on a diffuse network. Such technologies threaten to foreclose opportunities to realize thick community, but there still remains the possibility of steering toward more communitarian alternatives.

Community Energy
A close analog to indoor plumbing is the typical energy infrastructure in most electrified nations: massive centralized plants distribute electricity to individual homes. Energy users experience electricity as a commodity paid for monthly and commanded by flicking a switch or plugging into an outlet. The technology of centralized electricity distribution is, in the words of philosopher Albert Borgmann, a device.[9] It demands little engagement from the user and renders mostly invisible the mechanisms and social relationships through which energy is produced. The idea of experiencing electricity as a larger social unit is almost unimaginable within this energy paradigm.

Although there is no exact parallel to the communal well for electricity, energy systems could still be altered to support the experience of thick community. One way is to decrease the scale of centralization to that of a neighborhood, district, or town. Doing so would ensure that energy service boundaries more closely coincide with the reach of smaller-scale symbolic or social communities. One example is the district heating systems implemented by several European cities and towns. In Wildpoldsried, Germany wood-pellet-fired boilers installed in the basement of the town hall supply some forty-two local structures with heat, including all public and municipal buildings.[10] These systems, along with cooperatively owned solar panels or wind generators, serve to make energy production slightly more visible on a community scale as well as somewhat evoke an experience of community energy in the heating of public buildings.

Another option for enhancing the communality of energy systems is pursuing community ownership and governance. Such systems can be owned cooperatively, as is already the case for housing co-ops: Residents purchase "shares" in a nonprofit corporation, earn dividends, and vote on issues of maintenance and expansion. Though more communitarian than the consumer model of standard utilities, such arrangements still do not appear to strongly encourage interdependence or neighborhoods as a cohesive social unit.

Developing energy improvement districts (EID), based on the model of business- and block-level improvement districts, is a more promising pathway toward neighborhood- and street-level governance and ownership of energy, especially if set up as a microgrid.[11] Residents and businesses in an area would be bound into a governmental unit with the power to impose levies, build infrastructure, and regulate energy usage. The physical infrastructure would be "islanded" from the macrogrid. That is, the block or neighborhood would operate in energy networks much like a regional grid does today: local production from rooftop solar, wind generators, or other means would circulate within the microgrid before potentially being sold to other users. The neighborhood, in essence, would have its own energy infrastructure, including renewable energy storage devices, as well as a meter connecting it to the larger grid.

The EID model, along with a physical microgrid, would make the boundaries of the neighborhood- or block-scale social community more coincident with the practice of energy provision. Local electricity production could be managed as a common-pool resource much like a fishery or parking, thus providing a basis for local self-governance and economic interdependence.[12] Fees or taxes might be placed on users based on their energy usage above that which is locally produced. Those revenues could be used, in turn, to improve community energy infrastructure or to finance improvements to the energy efficiency of local buildings, including adding insulation or enhancing residents' ability to heat and cool their home passively. EID's could be integrated into larger energy districts in a way that combines local autonomy with regional coordination, as by neighborhoods electing energy officials at the district and city level to represent their interests. The communality of this arrangement could be further thickened by ensuring that important community-energy infrastructures are housed in a community building, which could be heated and cooled with locally produced energy and could act as a third place where residents come to discuss how to co-manage the neighborhood's energy infrastructure.

Local Economies and Retail Networks
Retail networks could be similarly relocalized. Local businesses serve several communitarian purposes: owners help cement and facilitate social relationships, provide mutual aid to neighbors and customers, and generally support a locale's economic vibrancy.[13] Local proprietors often also serve as "public characters."

The current dominance of regional malls, big-box stores, and Internet shopping draws citizens into retail networks that are simultaneously diffuse and massively centralized. A mall might serve hundreds of thousands of customers and be located in a sea of parking lots, hardly belonging to any particular neighborhood. Chain store managers face a clear conflict of interest between their allegiance to local community members and their responsibility to distant corporate leaders. Indeed, U.S. Supreme Court Justice William O. Douglas argued that the takeover of independent businesses and turning local proprietors into mere employees resulted in "a serious loss in citizenship" and the dilution of "local leadership."[14] It is, therefore, not at all surprising that economic concentration and the presence of Wal-Mart stores are associated with civic apathy and a sense of political inefficacy.[15] Consumers are framed as atomized individuals to an even greater extent when shopping online, where they are connected to vendors and distant goods by the massive networks owned by corporate entities like Amazon and Etsy.

Local retail, in contrast, contributes to thicker economic community. Studies show that local businesses recirculate several times the amount of dollars back into local economies that large chains do.[16] The lower *economic multiplier* provided by large chains is likely due to their greater reliance on other large, distant corporations for business-to-business exchanges and services and the fact that profits are invariably sent back to the home office. The greater recirculation of dollars by local retail means more dollars for other small proximate businesses and, hence, more jobs in the area. The more such businesses are located within any neighborhood, the greater the density and multiplexity of local economic relations.

Apart from the prototypical mom-and-pop store, a range of alternative retail infrastructures tend to be more communitarian than regional malls and online shopping. Consumer cooperatives, businesses owned by their share-holding customers, promise to encourage some loyalty to a local business over national chains through dividends and member discounts. There is, unsurprisingly, a lot of diversity. REI, a national sports equipment cooperative, facilitates outings between outdoor enthusiasts and contributes to local environmental conservation efforts. It remains a large chain store, however, located mostly in regional malls, and members participate only indirectly by casting a vote for the national governing board. Food co-operatives, on the other hand, are typically small scale, centering neighborhoods and downtowns, especially when they offer cooking classes and pursue other outreach activities. They generally emphasize sourcing from local farms as well, bringing urban dwellers in closer relationship to rural residents.

Another approach lies in substituting some retail activity with sharing. Within traditional thick communities, sharing was a natural outgrowth of the social, cultural, economic, and political structure. Hence, realizing it in already networked societies is challenging. Do-it-yourself (DIY) efforts by homeowners, for instance, often remain fairly individualistic endeavors: The DIYer buys or rents tools, obtains supplies from Home Depot, and seeks expertise in the form of YouTube videos or by contacting a member of his or her friendship network. Steering neighboring DIYers away from the practice of owning their own tools could be done through the tool libraries, which have been implemented in many American cities as well as in places like Kortrijk, Belgium. Tool libraries, when done well, not only provide a less expensive alternative to big-box tool rentals and purchases but provide a meeting place and mechanism for connecting DIYers with the social support and hands-on assistance of their neighbors. Tool libraries could be integrated into home repair cooperatives that take advantage of bulk purchasing opportunities as well as provide residents a space to store building materials.

Tools and home repair supply are far from the only goods and services that could be organized in this way. Child care has become increasingly expensive, especially where it is provided by large, for-profit firms. More cooperative arrangements lower costs, and thus partly alleviate the financial and temporal strains on parents that can discourage civic engagement,[17] while providing belonging to often isolated stay-at-home parents. Cooperative daycares would be partly run by stay-at-home parents and those with flexible work schedules, rotating shifts caring for each other's children under the supervision of one or more child care professionals. Cooperatively owned and run daycares realize the biggest cost savings when located in community spaces that are typically underutilized during works hours (e.g., churches), and, if properly designed, can serve as places where parents cultivate friendships and linger for gossip and small talk.

Such efforts can be, however, somewhat limited in their potential to create denser and more multistranded economic relationships because they connect neighbors primarily as consumers, sometimes to the detriment of nonconsumers. Consider recent efforts by management at an Albany, New York, food co-op to discourage unionization by its employees.[18] Food cooperatives, like other areas of retail, are often frequented by middle-class (sub)urbanites but staffed by poor minorities. Food cooperatives, however, need not be enclaves for affluent consumers; they could be more locally rooted and diverse communities by merging with a worker cooperative model. Employees could gain equity shares through their labor, represent themselves on the cooperative's board and develop both their professional and interpersonal skills, experiences that studies find to breed further civic engagement.[19]

One promising example is the Evergreen Cooperative, an employee-owned consortium of a laundry, a solar energy/weatherization company, and a greenhouse in Cleveland, Ohio. Workers become owners after a short probationary period, getting health benefits, a share of the profits and, eventually, thousands of dollars in assets invested in their workplace.[20] Though leadership there remains dominated by imported professionals and the path to achieving profitability has been bumpier than expected, the arrangement has permitted workers a great deal more political and economic power. Extending the Evergreen model to consumer cooperatives would permit workers to be included in their economic and political communities, not just consumers.

A similar tension between consumers and workers is present in the so-called collaborative economy: decentralized networks and apps, like Airbnb and Uber, that connect consumers looking for a place to stay or a ride with people renting spare rooms or offering an amateur taxi service. These arrangements better serve middle-class networked individuals than thick community. Airbnb rentiers have a competitive advantage over national chains like Holiday Inn and their working-class employees as much as they would for a hypothetical worker-owned cooperative hotel: guests often do not have to pay the same taxes and fees, and room renters have minimal overhead costs. Uber allows car owners to undercut taxi drivers, who are older, less educated, and more often non-white than Uber's workforce.[21] Members of the working class end up having their livelihoods partly undermined by middle- to upper-class Airbnbers and Uber drivers who might be merely supplementing an already living wage.

The demographic benefiting the most from these retail systems, however, is not cash-conscious middle-class consumers but those who control the networks. Indeed, Uber takes a 20 percent cut for their services in connecting drivers and riders, and Uber, not drivers, holds the power to set fares. Accounts of using collaborative economy applications like TaskRabbit, Fivver, and Mechanical Turk suggest that they are often used to extract poorly paid labor from those desperate for money without providing either benefits or labor protections.[22] For all their purported freedoms and flexibilities, work in the collaborative economy seems alienating. I cannot think of any work arrangement likely to feel more isolating than having one's livelihood be perpetually at the mercy of a distant Silicon Valley algorithm.

Such technologies could be steered toward creating more coherent, local economic communities rather than helping middle-class networked individuals find cheaper accommodations and Silicon Valley entrepreneurs "disrupt" the markets where members of the working-class earn their living. Amateur hoteliers and taxi drivers using these technologies could be integrated into local networks of worker cooperatives, institutions that would be enabled to levy fines on free-riding individualists, collectively purchase supplies, and pool funds to finance maintenance and provide benefits such as health insurance. Like energy infrastructures, supporting more communal retail systems would involve shifting ownership and control down from elite network supernodes to local groups, neighborhoods, and districts.

Transit and Transportation

Contemporary transportation networks tend to erect barriers to thick community, but differently than the infrastructures that I have examined thus far. Most obviously, they are about facilitating the movement of residents toward goods rather than the other way around. As I discuss in the previous chapter, automobile networks position drivers as individualized atoms shuttling from node to node through space that might as well be empty. North American transportation planning generally treats local streets as a mechanism for feeding cars onto limited-access highways and toward regional malls and business parks. Such road systems are geared toward getting people out of their neighborhoods instead of encouraging their engagement within them. Purposefully making traveling by car more inefficient and frustrating, therefore, would likely be a boon to local thick community.

At first glance it would seem that public transportation systems are hardly any different. Is not their function also to move bodies from one place to another? Differences are clearer if, as philosopher Robert Kirkman has advocated, one takes note of how "the physicality of the artifacts themselves plays a role, as do the physical capabilities and limits of the human body."[23] The fact that riders generally must walk to stations rather than simply exit their garage in a hermetically sealed automobile immediately alters the decision landscape presented by the local environment. Stopping at a store to chat with a friend becomes a simpler matter and going to more distant places is discouraged by physical fatigue or the trouble of having to make bus or tram transfers. On a limited access highway, in contrast, there is typically little physical disincentive to taking an exit a few miles farther away. A similar effect could no doubt be realized alongside automobility by eliminating free on-street parking or personal residential garages from locales. Indeed, regulations in Quartier Vauban, Germany, require residents in "car-free" areas to park their cars in lots at the edges of the neighborhood and walk home.

Regardless of whether it is accomplished through public transit or centralized residential parking, the point is that residents are encouraged to engage their neighborhood as pedestrians. The Jacobsian "ballet of the sidewalk" discussed in the previous chapter can only emerge if there are people on those sidewalks. Local cafés, pubs, and other businesses, moreover, have traditionally relied on foot traffic.

Given the likely continued dominance of the automobile for the foreseeable future, incremental gains could be realized by implementing a finer grained road network. Twentieth century ideas about the "proper" scale of neighborhoods and the perceived need to isolate cars to a small number of high-speed thoroughfares left most residential spaces without human traffic of any kind.[24] Therefore, rather than radically sequester the majority of automobile traffic to a select few roads, it is better to spread it out across pedestrian-friendly, traffic-calmed boulevards and one way streets. Early studies noted how those living on broad streets with heavy amounts of fast moving traffic knew fewer of their neighbors and more rarely spent their leisure time on front porches or sidewalks.[25] Limited-access, high-speed thoroughfares stymie walking and socializing. Indeed, many cities seeking to increase the sociability of their urban environments, including Portland, New York, San Francisco, and Seoul, have already replaced some of their elevated highways with boulevards, parks, and bike paths.

Also overlooked, at least in North America, is the potential to turn transit centers or district garages into multiuse community centers. Even bus stops can be pleasantly sociable "small urban spaces" if combined with ample seating, shade, and shops.[26] In Europe the intersection of two or more streetcar lines usually coincides with plazas or squares built to support lingering. The North American tendency to surround major transit stops with parking, in contrast, diminishes their potential as places for community sociability. There are, nevertheless, signs of that trend reversing. The government of the notably sprawling city of Calgary, Alberta, Canada, is planning to convert its Anderson light-rail park-and-ride into a mixed-use, pedestrian-oriented development. Regardless, similar to energy and retail networks, more communal transportation systems are finer grained and locate nodes and gathering places at the scale of blocks and neighborhoods.

Internet and Communication Networks

The current instantiation of the Internet is in many ways not unlike contemporary automobile networks. The metaphor of the "information superhighway" cuts both ways. It is tempting to see the Internet as even less supportive of local forms of community because distances are experientially distinguished by milliseconds rather than minutes or hours. Whether a piece of advice, a kind word, or an entertaining story comes from New Zealand or down the street has little impact on the user's online experience. Engagement with one's locality via the Internet is almost entirely voluntary. In fact, it is probably more accurate to say that it is discouraged, given that the greater diversity of goods, services, and contacts available globally means there is almost always something or someone more interesting or novel online than within one's neighborhood. Characters of only modest talent that residents might find at a local tavern or walking a downtown street must compete with the latest viral video to receive several million views.

Current Internet technologies are poorly compatible with the different dimensions of thick community. To begin, they help support a liberal moral order. One enters cyberspace as an individual and navigates networks as an atomistic client; it is an experience of nearly unencumbered mobility. Apart from passwords and other security measures, no legible relationship mediates between the user and the information they seek. Information comes as a free-floating and often costless commodity. Although web forums and other pages often become sites of social support and may be frequented by regulars, they face substantial barriers to thick communality. There is little incentive for them to be more than therapeutically contractual. Leaving is easy and commitment limited to providing a functioning email address. They are usually communities of limited liability. Barriers to multiplexity and systematicity likewise loom large. It is unusual for forum members to interact outside online environments, and online communities are typically not well integrated into larger communities, apart from the putative online "global village." Finally, recall the research showing that digitally mediated communication, for some people, is too thin to support a psychological sense of social connection.[27] Not everyone can experience social intimacy in online environments.

The communality of the Internet remains limited also because web forums, chat rooms, and social networking sites almost never function like political or economic communities. Sociologist Felicia Wu Song has found that users' and site owners' interactions reflect a consumer-proprietor relationship, hardly fostering political citizenship or social communion.[28]

Although spontaneous outpourings of aid do occur and some people turn their video game activities into business ventures, economic interdependence rooted in the needs of everyday life is more the exception than the rule in online community spaces. Philosopher of technology Darin Barney has argued that interactions premised primarily on communication and lacking a shared world of physical things and embodied practices constitutes an inevitably limited form of community. He contrasted online communication with the scenario of two neighbors silently flooding a local rink for a game of shinny hockey, contending that the physical engagements in the latter provide the grounds for a kind of communion rarely achieved in online environments.[29]

Internet technologies, therefore, would seem more likely to support thick communality whenever their physical infrastructure coincides with the physical scale of embodied social interaction. Consider Netville a Toronto suburban development that included a built-in high-speed local area network. Many of its residents used their network connections to cultivate neighborhood ties, not just surf the global information superhighway. The case Netville is often cited as demonstrating that the Internet writ large is not a threat to community life.[30] However, a more nuanced interpretation is made possible by noticing all the other factors that had encouraged the suburb's development of social community. Residents were connected to each other through an always-on, local-area network several orders of magnitude faster than the dial-up connections typically used to connect to the Internet at that time. Hence, certain high-speed features, like videochatting, were only feasible within the neighborhood and not with the broader Internet. Community resources were built into the local network, like a local email list and a library of reference materials. Moreover, it is hard to ignore the fact that residents were further bound together by being part of an ongoing sociotechnical experiment and later on by a collective fight to maintain their Internet service. Rather than supporting the claim that the Internet writ large poses no risks for community life, the example of Netville lends itself to a more modest conclusion: Internet technologies can be communitarian when they are designed to facilitate shared practices, collective resources, and the solving of political problems within localities.

Realizing Internets similar to or even more communitarian than Netville would require informational infrastructures to be more like community energy than a large-scale electrical grid. The current Internet is highly centralized and dominated by large service providers and giant corporations such as Google and Facebook. With high levels of centralization, it is difficult for the block, neighborhood, and district to have much salience. Proximity matters only to the extent that users just so happen to more frequently tweet or "friend" people within their metropolitan area or search for information on local events and issues than seek out niches in the global Web.

One promising avenue for a more communitarian Internet would be to promote wireless mesh networks, Wi-Fi systems developed not around centralized routers but emerging out of the linkages between local wireless computers and devices.[31] Indeed, studies suggest that involvement with community mesh networks promotes a sense of community and encourages civic engagement.[32] A mesh network not only creates a local-area network that can coincide with the boundaries of other dimensions of local thick community but also represents an opportunity for practicing politics.

Whether a mesh network can offer a sustainable alternative depends on its governance structure. In ad-hoc community mesh networks, access to the broader Internet is provided through users who voluntarily offer access through their ISP accounts. This laissez-faire model, however, makes it near impossible to monitor and govern bandwidth in order to prevent a "tragedy of the commons" situation, where the presence of too many freeriders renders the service unusable. Users who do pay monthly ISP fees, in response to such tragedies, would very likely close their connections to public use. A community governance model, as has been used in a mesh network in Wray, England,[33] enables members to treat Internet access as a common-pool resource. The potential advantages of such a model go beyond simply preventing freeridership. Pricing access so that Googling is more expensive than posting a question to the community's electronic bulletin board would encourage citizens to more often look down the street rather than to the other side of the world for needed information or for people sharing their interests. Going further, mesh networks could inexpensively host public versions of collaborative economy platforms similar to Uber. Such apps would help local economies cohere rather than "disrupt" them for the benefit of distant Silicon Valley venture capitalists.

Organizational Technologies

The same analysis can be applied to human infrastructures: organizations. Indeed, many organizational technologies frame the interactions between residents and providers of goods and services similarly to electric utilities or Internet companies. The organization of physical recreation, education, political action, and spirituality influences the extent to which daily life embeds people into thick communities. Each of these goods need not be organized as commodities for networked consumers. They could be offered in ways that are more coincident with the borders of neighborhoods and towns, governable by community members, and offering opportunities for building dense, multiplex, and systematic networks of social ties.

Physical Recreation

There are many different ways to organize physical recreation. Some efforts are completely asocial, such as exercising in front of a DVD of the latest at-home workout program. Only slightly more communal are the Planet Fitness or Gold's Gym franchises, where busy people pop in for a half-hour workout. Other fitness-minded citizens more explicitly seek out a sense of community via networked organizations like CrossFit or the November Project. CrossFitters participate either virtually by checking online for the "workout of the day" and posting their completion times or by joining CrossFit gyms to complete workouts in groups. Participants in the November Project use social media to arrange local meetups for collective workout sessions. Fitness fads, at least in North America, have increasingly become a dominant way for contemporary citizens to quest for community. Indeed, columnist Heather Havrilesky recently observed, "When I run on Sunday mornings, I pass seven packed, bustling fitness boutiques, and five nearly empty churches."[34]

Although organizations like CrossFit and the November Project certainly inject more social interaction and belonging into physical recreation, there is reason to question whether community is being mobilized mainly as a means to the end of individual fitness. Journalistic accounts of CrossFit culture finds that individual members are viewed as solely responsible for any injuries they sustain, even though the movement's "beat the clock" approach to weightlifting and cultural celebration of exceeding limits likely makes self-harm more probable.[35] CrossFit community apparently ends for a member once they injure themselves. Participants truant from a November Project meeting are publicly shamed online for "breaking a verbal." The website is rife with screeds ending with cultlike declarations of collective love for absent members and the hope that they will soon return.[36] Such declarations, however, look less like authentic acts of rhetorical communion and more like a strategy to keep members involved, similar to the "We Miss You!" letters the *New York Times* mails to the rare individual who has managed to successfully cancel a subscription.

Community physical recreation centers can often be much more communitarian than networked fitness programs. In fact, mid-twentieth-century American governments explicitly encouraged their growth in response
to worries about the anomic and atomizing effects of residential mobility.
Community recreation centers were sought as a procedural mechanism for
integrating new members into local communities.[37] The extent to which
such centers actually encourage belonging depends on their organizational
design and political structure. For instance, the Kroc Community Center
near Boston's Dudley Triangle, though an exemplar in economic communality in that local organizations and workers were involved in its design
and construction, is probably more accurately classified as a leisure center. Few locals in that disadvantaged neighborhood became permanent
employees, and residents entering the front door are greeted by a metal
detector—hardly a warm welcome. Its placement next to the commuter
rail line and the eighty-dollar-a-month family membership fees—which are
financially infeasible for many living in the area—further suggest that its
owner-operator, the Salvation Army, actually hoped to draw in suburbanites on the way home from their downtown jobs rather than community
members.

Community centers more deserving of the "community" designation
would substantively involve local residents in their operations, act in
ways that evoke and express trust, and serve as third places. Indeed, an
Ontario recreation center located within a similarly at-risk neighborhood
as the Kroc Center allowed residents to drive program development and
was found to be more than just a place to exercise but a lightning rod
for local social relationships and activity.[38] One program involved simply
providing a place for members to drop off their kids and chat with neighbors over coffee. Community-supportive leisure centers typically take on
the features of third places. Curling clubs, for instance, have traditionally
included their own bars, and many are designed to facilitate sociability.[39]
The Albany Curling Club, for instance, provides dues-paying members
drink tickets, which are used to buy the losing team conversation-starting
drinks after a match. Such third place features, however, are rarer among
sporting facilities today than previously. In the absence of such features,
sporting participants are deprived of the opportunity to cultivate a sense of
belonging in a local hangout. The emergence of a few yoga studios offering
classes at microbreweries and including a post-practice beer is suggestive
of the currently underrealized possibilities for greater community through
recreation.[40]

The organization of youth sports also affects the practice of thick community. The Little League system, for example, takes professional sports as its model. Youth are placed into coached teams where athletic performance is highly emphasized, frequently at the expense of the other possible social goods of sport.[41] Bureaucratically organizing youth sports may provide communal benefits to parents, of course, insofar as they are encouraged to interact when watching games or practices. It is a detriment to political community, however, in that children in such systems are deprived of the opportunity to organize their own play. Under the careful eye and handling of adults, children are unlikely to learn to negotiate their own conflicts.[42] Rather, they learn only how to follow rules and decisions handed to them by authority figures. Given that political community relies on not only the existence of suitable institutions but also people who can work through disputes constructively, it is undermined by the lack of such opportunities.

Schooling

Schooling probably contributes far more than the organization of sports to youths' development or failure to develop into capable communitarians. The typical American student is on the receiving end of some one thousand hours of instructional time each year, the equivalent of twelve months at a twenty-hour-a-week job.[43] Most schools' power structure leaves little room for teaching conflict negotiation and cooperation. The archetypal student competes as an individual against his or her peers to please an authority figure (i.e., the teacher) by performing requested tasks diligently and ideally with little complaint. The ubiquity of this power structure has led many educational analysts to criticize conventional, bureaucratic schooling as better at training citizens to accept authoritarian workplaces than at preparing them for the practice of democracy.[44] Such training not only contributes to ever more docile and acquiescent workforces but is likely to influence the practice of community more broadly. Increases in the rates of narcissistic personality traits and decreases in empathy and perspective taking among American college students since the 1970s suggests that, at a minimum, contemporary schooling does little to oppose advancing egocentricism.[45]

That is not to say that contemporary schooling is without communitarian benefits. Certainly to the extent that parents attend PTA meetings, fundraise for sports teams and music clubs, or serve on school boards, they are building denser and more multithreaded social ties within their locality. Nevertheless, similar to Little League, the organizational characteristics of most schools stymie youths' development into communitarian beings.

If schools were to be redesigned to aid the political development of youth, they would need to be less like bureaucratic organizations—some might say less like prisons—and more like communities. Developmental psychologist Peter Gray has outlined the benefits of democratic schools, such as the Sudbury model, wherein students codevelop the curriculum and help manage their own space.[46] The emphasis in such schools is on collaboration. Rather than compete with peers to perform better on standardized exams covering standardized curricula, students work with teachers to explore topics of their interest. Budgets, school rules, and the hiring and firing of staff are accomplished through town hall–style school meetings where students and staff have an equal vote. More schools could provide similar kinds of apprenticeship in shared governance, collaborative learning, and conflict negotiation.

Rather than insist on standard age groupings, the Sudbury model permits older children to aid younger ones with their learning. Studies of mentoring programs show that participants become more empathic as a result. Mixed-age play groups tend to be more cooperative and equitable. Similarly, the Roots of Empathy program, which aims to develop more empathic adults by having students interact and help care for a baby of a local mother, has been found to lessen aggressive behavior and promote prosocial action. Increasing the time that children spend caring for and teaching those younger than them is likely to produce less narcissistic and, hence, more communitarian adults.[47]

Schools could strike a better balance between the authoritarian-competitive model of good schooling and the principles of democracy and empathy. Doing so would likely entail fewer book problems and standardized tests and demand more hands-on learning and activities outside school walls. Realizing democratic schools, however, would require reversing much of the sociotechnical momentum that "modern" schooling has attained.

Spirituality and Religion

Not only could sport, recreation, and schooling be organized more communally but religion as well. It matters a great deal for community whether people meet their spiritual needs through watching the Evangelical "Hour of Power" on television, sporadically attending New Age retreats, or participating in local religious services. More and more people in technological societies are attending service less and less, seeking their enlightenment through purchasing audio CDs and following the blogs of spiritual gurus or—as is sometimes the case for the ostensibly irreligious—reverently following the metaphysical musings of high priests of technoscience like Stephen J. Hawking or Ray Kurzweil.

Often this shift in spirituality is presented as a turn away from "organized religion." A little reflection, however, leads to the realization that religious activities remain highly organized, albeit occurring more often via networks rather than in thick communities. As other scholars have illustrated, larger cultural patterns of religious individualism lead many people to approach different strains of religion or congregations not much differently than consumer goods like automobiles and kitchen appliances.[48] Involving oneself in any particular strain of spirituality is less often about reinforcing or finding a place in a local community. Instead, people primarily describe it as an individually therapeutic escape from the stresses of everyday life or a pathway to expressive self-actualization.

Running counter to the perceived decline of organized religion is the explosive growth of megachurches, which provide members a sense of community support and center their social activities but remain communally thin in several respects. As political scientist John Freie has described, many evangelical megachurches act as islands isolated from the wider region where they are located, substituting a flurry of internal activity for engagement with the outside world.[49] Indeed, some of the largest are sprawling campuses containing restaurants, gyms, housing, and a whole range of private alternatives to civic amenities. Many resemble malls or big-box stores, serving congregations numbering in the thousands and coming from miles away. Their utility in providing members with belonging, of course, should not be dismissed out of hand. Many, like the Saddleback Church in Lake Forest, California, organize members into small groups of a dozen or so people who meet and provide each other with social support throughout the week.[50] Although no doubt supporting the creation of networks of social bonds centered on faith, the tendency for megachurches to be both geographically and functionally disconnected from the surrounding locale limits their communitarian potential. To the extent that megachurches attempt to draw members away from involvement with outside social, economic, and cultural networks, they are the religious analog of exclusive gated neighborhoods—reflecting and reinforcing the fragmented social geography of suburbia.

Moreover, religious discourse within some megachurches appears rooted in an individualistic moral order. Many emphasize a personal rather than collective relationship with the divine and position community as "important only as a means of support for one's individual communion."[51] The outreach efforts of such churches often resemble marketing research, explicitly viewing potential members as spiritual consumers. Rock bands and variety acts supplement therapeutic, upbeat messages, and individuals' stories of struggle and redemption are used to create a multifaceted spiritual media experience seemingly modeled after television.[52] No doubt evangelical churches vary regarding the extent to which some of the above features are present. My point here is simply to show how certain relatively common design features better support individualism and enclavism than local social engagement.

Regardless of the sect, religious buildings and practices have traditionally served as social technologies for reinforcing community. Apart from their symbolic importance for the imagined community of a religious sect, churches, mosques, and temples can also act as social centers to local community life. Members come not merely to pray but to bond as well. The Christian ritual of communion, though sometimes seen as merely a symbolic act connecting individuals with Jesus, is simultaneously a moment of rhetorical communion for the congregation. Neither is baptism purely spiritual: it is a ritualization of the social adoption of a new community member. Postservice fellowship or visiting hours combine symbolic bonding with opportunities for talk and gossip; systems of offering or tithing bind members economically; and churches often center networks of mutual aid that, as the sociologist Eric Klinenberg showed in his study of the 1995 Chicago heat wave, support members and neighbors through times of crisis.[53]

Traditional religions, however, have tended to be weakest in political community. The Roman Catholic Church, as Dostoevsky's Prince Myshkin lamented, has historically taken the institutional form of an autocratic empire. On the other hand, sects like Presbyterianism are multitiered democracies: congregations select their pastors and elders, who go on to represent the church at higher levels, from presbyteries and synods to the general assembly. Religious organizations' practice of political community run the gamut from thin and authoritarian to thick and democratic.

Some religious institutions do considerably better than others in the degree to which they integrate themselves into larger social communities and in their thickness across multiple dimensions of communality. Traditional American black churches were a good illustration of this. They were as much centers of mutual aid, economic support, and political action as mechanisms for spiritual salvation. Such churches functioned as neighborhood social centers rather than enclaves, understanding their mission more through the language of social justice than personal evangelism and under-standing community as including those outside the congregation.[54]

Outside Evangelism and Mormonism, however, most mainline reli-gious sects are in significant decline, leaving future potentialities for thick-community-centering religious institutions in doubt—at least in North America and Europe.[55] Old churches are routinely repurposed into apart-ment complexes or, as was the fate of one Troy, New York, church, fra-ternity houses. The future of churches with rapidly aging congregations within societies increasingly skeptical of mainline religion certainly can look dire. Hope for the future may lie in their broadening the scope of their missions to dramatically increase community engagement. Religious institutions could enhance their role as centers of public togetherness by more often acting as omsbud-institutions for local sharing and mutual aid. Churches, mosques, and temples that tend to sit mostly vacant between holy days can provide inexpensive office space to community organiza-tions or local cooperative businesses. Religious leaders and congregations, moreover, might better integrate those doubtful of traditional metaphysics by elevating the rhetoric of social, economic, and political communion to the same level as the spiritual.

Indeed, citizens in the United States, Britain, and Australia have recently witnessed the emergence of Sunday Assemblies: church-like but nominally atheistic organizations with no set guiding dogma other than the "celebra-tion of life." Members of Sunday Assemblies sing songs together, listen to "sermons," and stay after service to mingle over coffee and cake.[56] Several of the communitarian techniques employed by traditional religious sects are clearly visible in Sunday Assemblies, albeit lacking the biblical, talmu-dic, or koranic language. At any rate, they appear to be filling a niche need among the increasing numbers of religious skeptics for a communal space to think with others about life's meaning. It remains unclear whether Sun-day Assemblies will be an atheistic analogy to the communal enclaves of evangelical megachurches or enhance a broader sense and practice of com-munity like the traditional American black church. Nevertheless, support-ing similar efforts to provide churchlike institutions to citizens may help provide the communitarian goods of traditional organized religions, even if the language and perceived metaphysical groundings are altered in the process.

Banking and Insurance

The same analysis concerning the organization of physical recreation and spirituality could be extended to a range of other goods. Frequently forgotten in the American debate between a private insurance system and nationalized health care is that mutual aid societies comprising national networks of local fraternal lodges used to provide sick and funeral benefits.[57] Too rarely is it considered that mutual aid societies could be a part of universal health care models. Rather than distributing insurance and regulating care through a national-level bureaucracy, a nested system of health insurance and care cooperatives could be subsidized by the state. Organizing health care in this way is not as infeasible as it might seem at first. For example, though poorer than much of the rest of India, the state of Kerala provides most of the country's palliative care through cooperative neighborhood networks supported mainly by local donations.[58] Moreover, Quebec funds a system of cooperative daycare facilities as part of its "social economy" regulations.[59] Parents only pay seven dollars a day per child and help steer the governance and operations of their local daycare. One could imagine a similar system for health care that would be driven largely by the interests of patients and doctors rather than CEOs, with equity encouraged via oversight by regional and national governments.

A similar analysis could be done for banking and finance as well as political parties and social movements. The discernible pattern in each case is a relative dearth of organization between the level of individuals and national or regional control. Social movements and political parties today rarely have local chapters, being more often networks of donors and magazine subscribers than thick political communities.[60] Likewise, there has been a 75 percent reduction in the number of banks with less than 100 million dollars in assets from the early 1990s to the 2010s; at the same time, the share held by the five largest banks has increased from 11 percent to 35 percent.[61] These larger banks not only charge fees so high that depositors are essentially paying banks to use their money for investments and loans, but they are also more likely to neglect the "social return" on their business—leaving riskier residents "unbanked" and awarding far fewer small business loans. For large corporate banks, it makes little sense to give risky loans for the sake of community improvement: their business is not in communities but only in networks of "bankable" individuals. Such institutions are effective at cementing local social networks and binding residents together in economic and political interdependence only to the extent that they are present and engaged at the scale of localities and neighborhoods.

Infrastructural networks and organizations are social technologies. They reflect and reinforce certain patterns of social relationships and practices. Technological civilization has increasingly moved toward networks and

organizations that are centralized at such high levels that they become
mostly incoherent to users, which often evokes a deceptive experience of
individual freedom. Users easily overlook the billions of dollars, reservoirs,
and massive treatment facilities that permit them to turn on indoor faucets
at their whim and receive water, and much the same is true of people using
an ATM, checking into a fitness center, or hopping on to a Wi-Fi network.
More recent retail innovations in the form of websites and apps like Uber
and Airbnbs connect consumers and producers of goods and services as
networked individuals rather than as community members. The aggregate
effect of these technologies is the structuring of everyday life to feel like
one is coasting across networks rather than embedded within webs of local
social ties and solidarities.

Insofar as large-scale networks characterize everyday life for citizens of
technological societies, individualistic ways of thinking are likely to dominate collectively held moral imaginaries. As network living becomes seen
as "natural," it shapes people's ideas about desirable ways of life. Indeed,
sociological studies have found that citizens increasingly define themselves
in individualistic terms.[62] Even politically active citizens have been found
to value civic engagement for its contributions to a sense of personal self-
actualization rather than its fostering of greater community solidarity and
well-being.[63]

Such ideas are not natural but are reflected and reinforced by sociotechnical systems, systems that are in turn supported and supplemented by
contemporary political regulations. As political scientist David Imbroscio
has noted, urban policy directed at the disadvantaged focuses on getting
individuals out of their neighborhoods rather than helping them improve
them.[64] In fact, one program was even called Move to Opportunity. The
oppression of poverty is seen as a result of limitations on individual mobility within residential networks—the ability to vote with one's feet—rather
than an inability to access the tools and resources needed to create more
stable and vibrant communities. In other words, such policies privilege
the ability to move over the right to stay put. Ironically, the effect of such
policies, despite however distasteful voters and politicians might claim to
find the "legislation of morality," is the imposition of a liberal, middle-class
moral order on disadvantaged populations.

Collective belief in the inherently liberating character of networks,
moreover, fails to recognize how power inevitably resides with those who
control them. A networked user can decide where to go on the network
but has little agency concerning either its geometry or its protocols. The
relationship between networked platforms like eBay and Facebook and

their users is like that of a Wal-Mart associate and their employer: the conditions of participation are set by the company, and those who do not meet those conditions are terminated from the network.[65] Workers on Uber or TaskRabbit, for instance, have little power to negotiate with those companies over their working conditions. They are little different from citizens who demand green energy from private utility monopolies or pedestrians trying to get around in cities dominated by limited-access highways. Indeed, Uber slashes fares and TaskRabbit dramatically alters the algorithm for connecting task workers with employers with no input or recourse for those trying to make a living through them.[66] Given these conditions, the common perception of distant networks as choice-enhancing boons to individual freedom seems problematic to say the least. Nevertheless, it is the moral imaginary increasingly instilled into members of technological civilization.

Making room for alternative moral orders and practices of community would require shifting at least some of the sociotechnical systems that structure everyday life toward enabling social interactions and exchanges. Services like electricity and Internet connectivity could be collectively governed as common-pool resources at the scale of neighborhoods and districts, enhancing political community as well as densifying local social ties. Opportunities for physical recreation and access to transportation and retail networks could better overlay with the scale of neighborhoods and city blocks, supporting talk and networks of exchange among neighbors. Important institutions like churches, banks, and interest groups, when appropriately scaled and integrated within community life, can help reinforce local social bonding, webs of mutual aid, economic interdependence, and political community. Designing infrastructural and organizational technologies in such ways would allow contemporary citizens to more often experience something like the village wells of Ibeica.

Table 5.1
Interconnections between Infrastructural and Organizational Technologies and Community

Dimension of Community	Thin/Barriers	Thick/Enablers
Social ties	Coarse-grained street design; peer-to-peer or highly centralized networks	Organization around community centers (e.g., religious, recreational, infrastructural)
Exchange/ support	Community used primarily as a means to something else (e.g., CrossFit, November Project)	Religious centers like traditional black church; tool libraries; local/cooperative business; systems for connecting neighbors (e.g., local mesh networks)
Talk	Online forums and "virtual" communities	Third place features (e.g., bars in recreation centers, religious fellowship hours); adequate and pleasant seating around transit centers
Symbolic/ psychological	Megachurches using consumer model of spirituality	District heating/municipal or community energy; infrastructures or organizations legible to users at local scales
Economic	Networked, peer-to-peer business (e.g., Uber, Airbnb)	Cooperative business; energy or Internet improvement districts
Political	Private monopoly of utilities; bureaucratically organized recreation; consumer-oriented virtual communities; mobility-oriented urban policy	Cooperative ownership and governance of infrastructures; self-organized play or community-developed recreation programs
Moral	Freedom as equivalent to existence as mobile, networked individual	Understanding self as embedded in and engaged with a web of relationships

6 Techniques and Gadgets: Socializing Individuals or Developing Communitarian Beings?

Techniques and gadgets are no less important than urban form, infrastructures, and organizations for developing thick community. Overemphasizing the latter technologies can give the impression that community is something that begins only when people walk out their front doors. Community, however, is not simply a public matter supported by large-scale technologies but is scaffolded on the sum total of everyday practice and private experience. Hence, I extend my analysis to the tools and techniques that help shape both public and domestic life. Gadgets, household appliances as well as approaches to child rearing, through their enablements and constraints, either stymie or spur along the development of communitarian beings.

Citizens cannot be expected to act in communal ways without experiences that teach them how to be communitarian. As historian Arthur Schlesinger observed, a central value of voluntary associations has been their role in apprenticing people in the practices of democratic political community:

> Rubbing minds as well as elbows, [association members] have been trained from youth to take common counsel, choose leaders, harmonize differences, and obey the expressed will of the majority. In mastering the associative way they have mastered the democratic way.[1]

How are the dispositions, beliefs, proclivities, and skills that enable citizens to act as thick communitarians rather than networked individualists influenced by the techniques and artifacts they use and have used on them? Techniques and artifacts, of course, cannot function without larger sociotechnical systems. A television is not worth much without communication networks to feed it with content. Nevertheless, the focus of this chapter is the question, How do small-scale technologies mediate citizens' psychosocial development? Do they encourage a retreat from thick community or stymie the development of thick social ties? Even though the use of techniques and things might seem insignificant at the level of individuals, I argue that their aggregate effect on the preconditions for thick community life can be substantial.

The Potential to Gather or Disperse: Domestic Technologies

I begin with an example that will likely seem irrelevant at first: home heating. Philosopher of technology Albert Borgmann rooted his analysis of the character of contemporary technological life on the distinction between a central heating system and the traditional wood-fired hearth, referring to the former as a *technological device* and the latter as a *focal thing*.[2] Both technologies, according to Borgmann, have a discernible patterning effect on the shape of everyday life. A central heating system is a device because it provides heat as a commodity to be called up on demand by setting a thermostat. The systems and mechanisms by which it produces heat are invisible to the user and, usually, wholly irrelevant to daily life. The hearth or wood stove is a focal thing because it demands that the user engage with its functioning. It cannot be stopped or started at will nor adjusted without some knowledge of building fires, ensuring proper airflow, and collecting and preparing wood. For Borgmann, the value of technologies like hearths lies in the meaningful focal practices that develop around them. The working and maintenance of a wood stove, much like craft work or motorcycle repair,[3] engages users' minds and bodies as it engages them with the material world. In contrast, central heating systems are simply turned off or on, and one places a call to an expert technician when they fail. Devices provide far fewer opportunities for focal practices.

The social dimension of this distinction is often overshadowed by Borgmann's analysis of the personal and experiential value of focal practices. The hearth, in his philosophy, does not merely support focal practices but serves as a center or focus for domestic activity as well. Keeping it running and collecting wood is often a cooperative activity among household members, and its warmth draws them together in social communion on cold days. One study elicited the recollection that "everybody sat in the living room because it was cold. And all sat around the fire, so we were more social."[4] Moreover, social psychological experiments have demonstrated that perceptions of physical and social warmth are tightly intertwined. Feelings of warmth, whether produced by holding a warm drink or being in a warm room, lead to more favorable evaluations of others, increased generosity, and more frequent use of relationally focused language.[5] The hearth is not merely a focus for domestic activities but converts the provision of physical warmth into a mechanism for social bonding. Many readers may have already noticed how social intimacy seems relatively easy around a bonfire or fireplace.

Central heating systems, on the other hand, promote the diffusion of activity throughout households. Indeed, they are probably more accurately termed *distributed heating systems*. The central unit is hidden in a basement, closet, or attic, and heat is distributed through a mostly invisible system of vents aimed at providing heating uniformly throughout a building. Not only are there few, if any, accompanying practices that support interaction, but central heating systems do little to physically incentivize social congregation. In fact, the implementation of distributed heating systems in public housing in England led to household members spending less time together.[6] Central heating, of course, does not prohibit household gatherings, but by rendering domestic community more voluntary, heating technologies can be seductions against it. The ostensible comforts of privatism often already appear as an alluring escape from the likely conflicts of collective living. Technologies like central heating act like a thumb on the scale, strengthening the appeal of physical separation.

The differences between distributed and nondistributed heating systems for community reflect and reinforce moral ideas about social reality. The Japanese variant of the hearth is the *kotatsu*: a charcoal or electric heater placed underneath a table-mounted blanket. On cold days in Japan's traditionally uninsulated homes, household members must gather around the kotatsu to be warm. This technology stems from and reinforces the Japanese cultural focus on relational collectivism. Indeed, as families in Fukuoka began "spending less time socializing and more time in individual activities" in the late twentieth century, was it merely a coincidence that they were heating more rooms within their houses?[7] Thus, the design of home heating is no mere technical decision but a mechanism for strengthening certain social values and practices at the expense of others.

Distributed central heating also reflects and reproduces a networked individualistic moral order. One study framed the shift from central hearths to distributed heating as leading to an ostensibly beneficial reduction in intrafamilial conflict as children then spent more time in their own bedrooms.[8] This framing reflects the liberal moral injunction of private suburban life: minimize conflict through the avoidance of others. To be fair, a wood or gas stove, kotatsu, or other focal heating unit does not force household members to resolve their conflicts productively—they may still choose to ignore them or act childishly. Because such systems physically incentivize copresence, however, they discourage isolation as a solution to conflict. In contrast, distributed central heating systems provide an ersatz version of domestic harmony, as does suburbia on a larger scale, by limiting the possibility of interpersonal conflict and thereby interfering with the development of the negotiation skills needed for political community.

Despite early hopes that televisions or family desktop computers could become "electronic hearths," such technologies' ability to center domestic life turns out to have been more mixed. Television viewing no doubt does not always mean glazed-eyed passivity, because two or more viewers can talk over and around their televisions. Potential interaction is limited, however, by the increasing prevalence of TVs in bedrooms: Children with their own set on average spend roughly 20 percent more of their time watching television alone.[9] As is the case with distributed heating systems, bedroom TV sets allow family members to more easily sidestep conflicts and promote social diffusion. In fact, even Dr. Spock advocated them as a way of minimizing parent-child conflict over what to watch.[10] The consequences of shifting from the family computer to laptops and tablets is probably much the same: social diffusion and the avoidance of conflict.

At the same time, would one really want to equate the social interactions happening around a kotatsu or wood stove with those between people watching the same television or computer? In the latter case, participants' eyes are more often drawn to the screen rather than each other's faces, and the conversation material is more frequently dominated by that night's programming or the latest viral videos. Even though the continued presence of some form of talking makes overly dystopian portrayals of mass media devices as turning viewers into antisocial zombies seem hyperbolic, it is reasonable to worry about potential declines in the *quality* of conversation. Political scientist Janet Flammang points out that "conversation is an art best learned through an apprenticeship with skilled conversationalists."[11] Good conversation requires the ability to introduce and talk about controversial topics without being alienating, to pay attention to the needs and interests of others, and to artfully tell stories. Good conversationalists are capable of reading the emotional states of their interlocutors by observing body language and facial expressions.[12] Even though people can bond over watching *Breaking Bad* or cat videos on YouTube, rarely is such bonding either a product of good conversation or an apprenticeship in talking.

Learning the skills of civil conversation and bonding forms of talk requires appropriate technological arrangements. Are digital devices up to the task? As one teenage boy who primarily communicated with his friends via texting told Internet scholar Sherry Turkle, "someday, someday ... I'd like to learn how to have a conversation."[13] This state of affairs among youth in technological societies contrasts sharply with the social precociousness of children in communities that lack televisions and other media devices, such as those in traditional New Guinean villages.[14]

In addition to limiting screen time, strengthening conversational skills can come from increasing opportunities for talk. Flammang describes the communal and family dinner table as a technology around which the practices and skills of thoughtful, civil conversation can be cultivated.[15] Partly for this reason, the Slow Food movement opposes the stop-and-refuel eating practices typically associated with fast food and microwave dinners, hoping to maintain the traditional European practice of long mealtimes. For households facing a time crunch from their harried lifestyles or allured by the conveniences of individual microwaveable dinners, however, communal dining can feel more like a chore than a pleasure. Deficits in the requisite time, money, and expertise may make it more difficult to pursue a slower, talk-based approach to mealtimes.

Given the influence of the above domestic technologies, the growing proclivity against living with others or making relational commitments, found by sociologists such as Eric Klinenberg,[16] seems understandable. People who grow up in environments that enable and encourage social diffusion and solitary leisure through personal technologies are less prepared to productively work through domestic conflicts with roommates or romantic partners. My point here is not that previous generations did not face their own barriers to constructive conflict resolution. Rather, I mean to suggest only that such technologically shaped experiences work against all the prosocial instruction contemporary citizens might receive elsewhere, including in therapy sessions or through the workplace. At any rate, inexperience with conflict negotiation and emotionally attentive face-to-face talking has ramifying consequences for community. Indeed, social movements guided by a "personalistic" ethos—where members' participated primarily for the gains to a sense of individual self-actualization—struggled to act as a collective and tended to disintegrate when faced with conflict.[17] Recall the similar tendency, among suburban gated community and high-rise-housing cooperative residents to move rather than directly confront contention. This behavior only seems normal when it mirrors peoples' early formative experiences with conflict: avoidance. Would such a moral order be likely to change without parallel shifts in domestic technology?

Domestic environments help cultivate a strong, arguably overdeveloped, sense of individuality. The heating and personal technologies that promote domestic social diffusion work similarly to the networks and organizations discussed in chapter 5. The fact that they make their goods immediately available for the individual user and with little engagement from other household members helps to frame households not as small-scale communities but as systems of nodes for the flows and movements of networked individuals, with parents sometimes ending up feeling like taxi services for their children, or family members having to e-mail each other to coordinate their schedules.[18] Much of citizens' preparation to become operators of social networks rather than thick communitarians begins in the home.

Technologies of Child Rearing

Domestic technologies do not give birth to networked individualism by themselves but parallel and strengthen understandings of social reality imparted during childhood and adolescence. The predominant approach to establishing infants' sleep patterns in North America and Europe remains the cry-it-out method, wherein infants are placed in their cribs, often in separate rooms, and parents refrain from coming to comfort their child for varying lengths of time. Some even advocate "going cold turkey." The process is meant to help infants learn to "self-soothe." Although this practice might seem perfectly natural to many Americans, it is a cultural peculiarity. Both prior to the twentieth century and around the world the standard practice has been the immediate soothing of infants by caretakers as well as bed sharing.[19]

The emergence of solitary infant sleeping and "crying it out" as the prevailing wisdom came with justifications that it better instilled self-reliance and independence. Needing physical contact with another human being in order to sleep was, in turn, framed as an unhealthy form of dependence, somehow different from the myriad other ways young children are utterly reliant on others. Cultivating this kind of "self-reliance" has become valuable, as anthropologist Eyal Ben-Ari has noted, because it speeds the integration of children into the work schedules of parents in industrialized nations.[20] Cosleeping or frequently waking to tend to a fussy child is rarely compatible with an inflexible nine-to-five workday or the requirement that people work forty hours or more per week. Independently sleeping infants are valuable because most parents have little autonomy regarding their own sleep patterns. At the same time, one should not discount the benefits to parents from the hour or so of "child-free" time that such practices can ensure, especially given how many technological societies render stay-at-home parenting a full-time and largely solitary endeavor. Regardless, crying it out helps naturalize the moral imaginary of the unencumbered self. It is a psychocultural technology that lays the groundwork for the "therapeutic individualism" described by Bellah and his coauthors.[21] It trains children to think of themselves as mainly responsible for their own sleep; too much dependence on parents is framed as potentially pathological.

Similar moral ideas undergirded the twentieth-century shift toward most children having their own separate bedrooms. Early psychologists advocated for private bedrooms not only as a means to cultivate "self-reliance" but also because of Freudian-inspired fears of "momism," the pathological dependence on one's mother, and worries that room or bed sharing encouraged incest and threatened children's "sexual hygiene." Such arguments gradually gave way to the belief that private space is a stepping stone to adulthood and a necessary part of an individual's expressive development. The percentage of teens with their own room in the United States increased from some 30 percent to more than 80, probably playing some role in the near tripling of the average home size over the twentieth century.[22] However desirable or pleasant one's own space might be, providing teenagers with a surfeit of solitary space is likely to produce adults that expect a level of privacy only found in suburbia or solo dwelling. Would it be reasonable to expect youth to grow up to be capable communitarians when the increasingly prevalent adolescent rite of passage is the provision of domestic isolation?

The private bedroom, moreover, plays a significant role in enculturating youth to be consumers. A private bedroom becomes a space to be filled with things reflecting one's individualized identity as a budding consumer-adult. Photographer James Mollison's collection of photographs of "Where Children Sleep" from across the world illustrates the considerable effort and expense to which affluent Western children go to materialize their identities.[23] An eleven-year old American hunting enthusiast has a bedroom festooned in "camo" patterns and weapons, and a fourteen-year old Scottish girl drapes her room in photos of rock stars, police caution tape, and skull-and-crossbones flags.

French social theorist Jean Baudrillard described this practice as the curating of the self through collections of objects, substituting conflictual human relationships with a feeling of consumer freedom and purchased uniqueness via a personalized relationship with things.[24] Teenagers in consumer-oriented technological societies learn to retreat from domestic relational conflict to the comfort of systems of consumer goods that reflect and reinforce the unique personal identities they have sought to buy for themselves. No doubt bedrooms can just as easily be filled with gifts from others that reference and support relationships,[25] but such objects must compete with alluring consumer products. As consumer culture infiltrates the lives of youth, buying and having is put into conflict with the practices of being and relating—which, in turn, upsets parent-child interactions and negatively affects children's well-being.[26] The continued effect of consumerism into adulthood has been depicted in the movie *Fight Club*, as the protagonist flips through an IKEA catalog on his toilet while describing how his carefully designed apartment reflects who he is or, more likely, who he wishes to be.[27]

Regardless, the belief that separate sleeping arrangements and crying it out produces more self-reliant, confident adults has never been borne out in observation. In fact, geographer Jared Diamond described the cosleeping youth of Papua New Guinea as reliably developing into capable adults without any signs of pathological forms of reliance.[28] Some developmental studies have found that cosleeping children tend to be emotionally healthier and even show certain forms of autonomy earlier than solitary sleepers, in particular dressing themselves and making friends independently.[29] Other studies, however, have found no significant relationship between sleeping patterns and pathological behavior.[30] Given the lack of firm data that crying it out and separate bedrooms actually produce better functioning people, might it be that their function is to adapt infants to "modern" schedules and provide parents a much needed respite from the stresses of childcare, among other practical concerns?

Solitary sleeping and crying it out, in any case, set the stage for naturalizing individualistic understandings of independence later in life. It frames autonomy as achieved through the ability to disconnect rather than integrate. Consider how, especially in North America, England, and Australia, young children have become increasingly sheltered within the home or under the close eye of parents. Their mobility within their towns and neighborhoods has shrunk over the course of a couple of generations from several miles to as little as no further than the house next door. A case in point is how the percentage of British seven- and eight-year-olds allowed to walk to school unsupervised fell from 80 percent to less than 10 percent between 1970 and 1990, with similar trends existing in countries like Australia and the United States.[31] More and more children, therefore, experience autonomy as a freedom achievable only by disconnecting from their parents by surfing online spaces or driving a car. Outside these spaces, autonomy for Western youth is exercised mainly in the cultivation of an individual expressive identity, which often amounts to purchasing certain kinds of music, clothing, room decor, and other paraphernalia. Freedom under such terms equals the individual operation of communication, retail, or transportation networks.

Youth in North America and Europe as well as in most other parts of the world have traditionally enjoyed a great deal of geographic mobility. Jared Diamond, for instance, recalled how a ten-year-old New Guinean boy was allowed to leave his home for over a month to help move equipment from one village to another.[32] The kinds of autonomy provided by an automobile or an Internet connection and that experienced by Diamond's young assistant are quite different. In the latter case, geographic mobility was allowed because of the careful eye of neighboring villagers. In much the same way, family friends, neighbors, and acquaintances would watch out for children walking to school in the early to mid-twentieth century United States.

Child-rearing approaches thus differ with respect to how they frame independence: Is it a product of individual atomism or community embeddedness? The feasibility of any given approach is no doubt a collective issue. It would be unreasonable to ask parents to give their children a long leash in an acommunal neighborhood. That would leave them vulnerable. On the other hand, intensive parental supervision and low levels of community engagement form a vicious cycle: sequestered children cannot cultivate friendships that draw neighboring parents together or build social trust. In turn, a lack of trust motivates parents to keep a short leash on their children. The retreat of childhood into the home, or to highly structured, adult-supervised hobbies, reflects and reinforces a dearth of local social connection. Indeed, as several recent cases well demonstrate, it seems as if many people are willing to do no more to help coparent local children than call the police and have parents arrested when their kids are found playing without supervision in a public park, left in a car for five minutes during an errand, or discovered "playing hooky" from church.[33] The culture of avoidance and lack of care increasingly characteristic of contemporary neighborhoods leads to the state taking over functions previously accomplished by community. Anonymous calls to policing institutions come to replace the soft surveillance, nosiness, and care enacted by neighbors in ensuring the collective safety of children.

The parenting practices that have created what Karen Malone has termed the "bubble-wrap generation" have still more consequences for community.[34] The replacement of unstructured play with formalized sports and classes means that parents spend more time shuttling children around and inevitably less time maintaining their other social ties. Parenting can quickly become more of a retreat from community than a further embedding into it. Moreover, as I noted earlier, overly structured environments decrease the number of opportunities for youth to learn to cultivate and organize their own communities.

Acommunal living areas are unlikely to develop flourishing social networks overnight, and perhaps conditions are wrong for already-built suburbs to turn into seven-dimensional communities. Nevertheless, more communitarian forms of youth autonomy could be established by setting aside more spaces for unstructured play. Consider the Land, an "adventure playground" in Wales that provides children with access to all sorts of building materials, hand tools, barrels to start fires in, and a creek. Although they are supervised by playworkers, the latter only intervene when an accident looks to be imminent.[35] Such playgrounds provide the support, and a minimal level of safety, for children to play more collaboratively, test their own limits with regard to safety, and teach their younger brethren how to start fires and build forts. Although some degree of self-organization can occur in traditional playgrounds, the way that children can collaboratively build and unbuild their play environment in spaces like the Land provides greater potential for the development of political community and a psychological sense of place. Adventure playgrounds offer children the ability to enact their own versions of the communal barn raising. The involvement of local residents in the provision of tools and building materials at the Land, moreover, suggests that adventure playgrounds have spill-over benefits for intergenerational communality. Regardless, such semisupervised spaces help open the door for children to more often realize autonomy via community integration rather than networked disconnection.

Techno-Cocooning

Gadgets, tools, and devices may also limit the potential for thick community by discouraging public involvement and sociability. Information and communications technologies (ICTs) can displace other forms of interaction in everyday life. For example, "More television watching means less of virtually every form of civic participation and social involvement."[36] TV watching partly displaces thick communality with weak symbolic communities knitted together by shared viewing. Part of the displacement effect is rooted in the development of *parasocial* relationships with television characters. Viewers have been found to experience a sense of belonging when watching favored TV shows, despite the lack of actual reciprocated interaction.[37] Insofar as this sense of community partly satiates the felt loneliness of viewers, especially for stay-at-home parents or the unemployed, it disincentivizes broader community involvement. It would be unfair to solely blame television for this outcome, however; the relative dearth of vibrant public spaces, at least in areas like North America, leaves citizens with little alternative.

Media-driven domestic cocooning is exacerbated by the degree to which citizens can afford and pursue increasingly sophisticated home theater systems as well as streaming services offering movies and television shows on demand. To be fair, there remains little communality around most cinemas left to displace, given their movement out of neighborhoods and into regional mall multiplexes surrounded by a quarter mile of parking lot in every direction. Nevertheless, the multiplying conveniences and richness of these technologies persuade and seduce users toward making the home more and more the site of their leisure time. Further driving this shift are increased anxieties about public danger. Media scholar Barbara Klinger has described how home theater systems are marketed as "providing self-sufficiency and refuge from the hazards of the public sphere."[38] Although Klinger was mainly discussing the perceived hazards presented by terrorism in a post-9/11 world, it would be surprising if contemporary anxieties over mass shootings and other crimes did not motivate similar purchases. Indeed, some observers have associated the sizable uptick in purchases of high-resolution televisions and high-end cable subscriptions to the growing perception of public sporting events as ever more dangerous—a likely contributor to recent declines in professional sports attendance.[39]

It is unlikely that television, especially home theaters, could support thick community. Research into making television more "social" tends to focus on seamlessly integrating texting and social media into the viewing process rather than building local solidarities.[40] Though perhaps helpful for networked individualists wanting to simulate the collective watching experience without the obligations that come with playing host to other human bodies, such technological developments would do little to get people to leave their dens. In any case, one of the few areas where something like communal TV watching could be realized is televised sports. Even the smallest European towns will host public viewings of popular soccer matches on large outdoor screens. Although such practices directly support relatively thin forms of symbolic community via collective spectacle, the potential for talk and establishing local social bonds means they are still an incremental improvement over solitary viewing. In any case, it may take fairly drastic measures to end television viewing's dominance of citizens' leisure time in industrialized nations, ranging from around a quarter of people's free time in Germany to over half in the United States.[41]

Technologies like air conditioning are little different than television in their communal effects. In contrast, front porches, plazas, and public water fountains are all urban design features that help provide a respite from the summer heat in ways that draw residents out into public or, at least, semi-public spaces. Air conditioning, though admittedly pleasant and convenient, is a material incentive to stay indoors. Indeed, the presence of A/C discourages residents from occupying their front porches and, hence, they are more seldom used for neighboring.[42] Front porches, moreover, have been decreasing in size under the pressure of expanding garages and driveways to an extent that they have become more symbolic than functional. Their usage is further discouraged by relatively empty suburban sidewalks and the draw of backyard pools and decks made alluring by increased cultural expectations for privacy. Such changes exert a negative force on public communality. Political scientist Eric J. Oliver explained part of the relative dearth of civic engagement in newer Sun Belt cities, such as Phoenix, with their greater reliance on air conditioning and lack of large porches and shaded sidewalks. Privatized cooling technologies encourage more privatized forms of sociability.[43]

The relationship of digital ICTs with thick community is more complex: they give as well as take away. E-mail, social media, and cell phones are all social technologies in a way that television is not. Still, they enable the supplanting of local, thick community with networked individualism. Similar to how TV offers surrogate parasocial relationships, the extent to which e-mail and online social networking make remaining homebound less lonesome persuades people away from public communality. ICTs are admittedly little different from the telephone in that regard, and this, of course, can only be true for those who are not cyberasocials. Regardless, any improvement in people's ability to maintain distant connections inevitably means some displacement of more proximate forms of socializing. People have only so much free time, so much need for social intimacy, and can maintain only so many social connections. Calling distant contacts or texting a friend across town are alluring options when faced with the uncertainties inherent in establishing deeper relationships with nearby coworkers or popping in next door. Long-distance communication devices enable users to rely on already established and less proximate bonds and avoid the risks and effort involved in getting to know neighbors and strangers. It is therefore hardly surprising that sociological studies of one Toronto neighborhood found that the frequency of face-to-face interactions between close neighbors had decreased to one half their 1978 levels by 2005.[44] ICTs help drive networked individualism.

The degree to which Internet access writ large displaces rather than facilitates local community engagement partly depends on how it is used and by whom. Social butterflies are typically as active online as off. Those who are satisfied with and participate in their geographic communities pursue more community-oriented uses of ICTs. Furthermore, given ICTs' logistical advantages over the telephone and letter writing, it is not at all surprising that the most civically engaged of citizens usually happen to be avid e-mail writers and online social networkers.[45] Yet, ICT use just as often comes at the cost of communal activity. Some studies have associated an increased use of ICTs with increased time spent at home, greater feelings of loneliness, and less time talking face to face with friends and family.[46] Others show that, although social ICT use can engender feelings of connection, a sense of isolation often remains. Users who utilize ICTs as a means for coping with loneliness and poor social skills are liable to get stuck in a vicious cycle: compulsive Internet use leads to negative life outcomes like missing work or social engagements and, hence, greater feelings of loneliness.[47] It appears that some people's pursuit of Internet-enabled connectedness provides weak or negative social returns. So contemporary ICTs are communally ambiguous at best: motivated communitarians and those complementing already extensive offline social networks probably benefit from them, but the chronically lonely, the socially unskilled, cyberasocials, and those discontented by deficits of face-to-face talking often do not.

I do not mean, however, to downplay the importance of online spaces as refuges for those excluded from their local community because of disability, sexual orientation, or gender identity. Philosopher of technology Andrew Feenberg, for instance, argued that the existence of online forums for those left relatively homebound by Lou Gehrig's disease, to take one example, illustrates the communal potential of online spaces, namely offering disadvantaged populations the chance to discuss their worries and tribulations with people like them.[48] Similar references to gay or transgender teens living in intolerant areas are practically cliché in discussions of the Internet and community, though not without good reason. Virtual gathering spaces for those marginalized within their local environs no doubt provide them with a much needed sense of belonging and relative safety.

On the other hand, such spaces seem less liberatory once one recognizes their similarities to the urban ghettos that have housed and provided rich community life to religious and racial minorities throughout the centuries.[49] Few would extol the liberating potential of ethnic ghettos. I see little reason to get too excited over virtual ghettos being made available to people otherwise excluded by contemporary technological societies. Would it not be more just for all citizens, including the aged, disabled, and sexual/racial minorities, to be able to realize multifaceted forms of communal belonging rather than be sequestered to virtual social networks? The provision of quasi-ghettoized online communities for the oppressed and collectively forgotten but thicker forms of community for everyone else would hardly be a desirable outcome with respect to principles of social justice or communitarianism.

At the same time, the example of marginalized and oppressed groups finding some sense of communal solidarity online is often used to imply that worries about digital devices and isolation are overwrought or a concern of the "privileged" (and hence are ostensibly unimportant). Such arguments, however, fail to give compassionate recognition to people's diverse subjective experiences. For those who find digitally mediated social interactions less fulfilling, a society increasingly predicated on them can feel alienating, if not oppressive. In the same way that a socially just American society would recognize that African Americans and other minorities do not experience the criminal justice system or the labor market in the same way as whites and attempt to rectify the resulting inequities, a technologically just society would admit that people experience sociotechnical arrangements differently and seek to limit the extent to which harm is disproportionately shouldered by certain populations.

Although I have been mostly discussing Internet use through a home-bound computer, the same conclusions apply to portable digital devices. Cell phones allow users to constantly remain in contact with their already established strong ties. They enable *tele-cocooning* or *social privatism*. Much like the personal automobile, portable ICTs allow their users to bring a modicum of private space with them wherever they go. Indeed, mobile phone users, and to a slightly lesser extent those on laptops and tablets, are notably inattentive to their surroundings in public and semipublic spaces—much more so than book readers. ICT users often adopt a physical stance that closes off interaction, avoids the gaze of others, and disrupts copresent conversation.[50]

The extent to which users are hunkered down with their devices of course depends partly on their intentions. For those explicitly seeking public communality, their device is often merely an excuse to be out at a Wi-Fi hotspot rather than a shield used to block contact with other people.[51] However, the choice between public communality and tele-cocooning is hardly an unbiased one. Mobile devices frequently seduce users into privatism because they offer an alluring escape from the risks inherent in interacting with strangers or mere acquaintances. A text to one's spouse or partner is a much less anxiety-provoking way to pass five minutes waiting for the bus than talking to a nearby person. San Francisco secondary students assigned to go through a three-day digital detox, for example, learned that, without their devices, they more often conversed with family members and interacted with people they usually did not.[52] Professional sport coaches, moreover, have complained about declines in team bonding as players increasingly cocoon with their smartphone or mp3 player when in dressing rooms or on the team bus.[53] Again, the satiation of the need for belonging via distant ties often comes at the expense of more proximate interactions.

Several recently developed technologies go much further in depressing talk and public sociability. Smartphone apps, like Cloak and Split, enable users to track their contacts in order to avoid running into them in public. Restaurants like Panera Bread and Chili's are replacing their waitstaff with touchscreen ordering kiosks.[54] These are more explicitly antisocial technologies. The former enhance social privatism and enable users to shun potentially conflictual or boring interactions with ex-partners or acquaintances. The latter, though replacing an interaction that is already to some degree socially automated, nevertheless results in a further decrease in opportunities to engage in embodied face-to-face conversation with another human being. Self-ordering kiosks are unlikely to have a button labeled "small talk." While those with already vibrant social networks might look favorably on the ostensible increases in "efficiency," for others, the loss of such routine encounters may eliminate a sizable chunk of their everyday embodied social contact.

Although some network sociologists depict many of the technologically enabled capacities discussed above as simply a desirable expansion of choice regarding social interaction,[55] such a view relies on the presumption that people are choosing—or even can choose—their social contact rationally, in line with their own reflective preferences. It assumes that people are actually content with their "choices" to "check out" of public sociability. The reason that television and other media devices evoke such ire is that many users recognize that the emotional and psychological pull of those devices leads them to develop habits that they do not find desirable or healthy, upon reflection. Indeed, psychological studies suggest that many people actually do not prefer digitally mediated activities but choose them in part because of the lower amounts of risk and effort they demand.[56]

As Sherry Turkle has argued, "Technology is seductive when what it offers meets our human vulnerabilities."[57] I suspect that most readers have regretted letting a marathon of Netflix viewing eclipse a night out, or pretending to read text messages when alone in public rather than striking up a conversation. Although ICTs and other media are certainly used by some to pursue greater community, much of that potential is left to users' individual cognitive resolve as well as their underlying social aptitude and disposition. Social belonging becomes more a personal responsibility than a public good. Thus the communitarian compatibility of current ICTs seems inevitably limited. Although they enable social connection through networked individualism, those who prefer the kind of face-to-face talking and public sociability characteristic of thick community have good reason to desire very different ICTs or restrictions on their use.

Technological "Retreats" from Community

Other technologies are much less ambiguous in their communitarian consequences. Consider concealed-carry handgun licenses (CHLs). Though firearms no doubt can support forms of thick community, namely within the shooting and hunting clubs that serve as third places and center local social bonding, concealed-carry handguns signify and reinforce a symbolic, moral, and practical escape from local thick community. Sociologist Angela Stroud found that many Texan CHL holders ascribed to hyperindividualistic and antisocial moral imaginaries.[58] Citing potential victimization as a reason for obtaining a CHL, her interviewees acted in ways that reflected an intense fear of the world outside their door. They tried to avoid any public contact with strangers, some even bringing a loaded gun when answering the front door. The technology of the concealed/self-defense handgun appears to enable and reinforce a more tense, fearful, and antisocial approach to public interactions, at least in some populations. Indeed, armed motorists are more likely to engage in hostile "road rage" behavior, and others have well described the increased recklessness that can come with owning a handgun.[59]

One illustrative example occurred when a Kalamazoo, Michigan, CHL holder, Walt Wawra, visited Calgary, Canada, albeit without his concealed firearm.[60] Ostensibly feeling accosted by two Canadian men in Nose Hill Park asking him if he had been to the city's annual rodeo event yet, he placed himself between them and his wife and abruptly ended the interaction. Wawra's ensuing letter to the *Calgary Herald*, complaining about Canada's stringent handgun laws and his gratitude that these two strangers had not pulled a weapon on him, amused Canadian readers. To them, Wawra appeared to suffer from a pathological form of paranoia: Nose Hill Park is not particularly dangerous, nor are public conversations, at least in Calgary, something to be frightened of. In light of Stroud's research, Wawra's exaggerated fearfulness to seemingly mundane and innocuous public interactions appears to be common among many CHL holders. The technology, of course, does not do this by itself. An overdeveloped sense of "stranger danger" is amplified by CHL training, which depicts public spaces and interactions as inherently threatening.

Many of Stroud's interviewees, moreover, associated having a CHL with "self-reliance" and not having to "depend on society," which should be a familiar acommunal refrain to readers by now.[61] Deriding the moral resolve of people who rely on police for protection, Stroud's interviewees championed a personal responsibility model of security, seeing themselves as more exemplary citizens and ideally unfettered individuals. This extreme individualism was also apparent in their attraction to "survival prepping." Many of the CHL holders interviewed by Stroud regularly stored caches of food, water, and ammunition, even going so far as to purchase rural land to prepare for the possibility of societal collapse. The Hobbesian and Social Darwinian subtext is clear: Stroud's interlocutors perceived American society to be at the precipice of a war of all against all, within which only the most self-reliant, well-armed, and prepared—the fittest—would persevere. In fact, one person even described such a scenario as "God's way of thinning the herd."[62]

Regardless, turning to CHLs out of the fear of victimization, ironically during a period of historically low crime rates, undermines the potential for thick community. It signals that citizens have given up on collectively achieving security, preferring a private technological fix. To be fair, there is a kernel of reasonableness to citizens' desires for CHLs when it is a response to perceptions of police inefficacy. Police do not typically arrive on the scene while the crime is still in progress, and for some populations, officers are as often victimizers as public servants.[63] Concealed carry might feel like the only reasonable option when police departments are increasingly distant, centralized, and bureaucratic and when informal community-level mechanisms for ensuring safety have eroded. Policing is rarely subjected to local public oversight, and officers are too frequently cocooned within their patrol cars. Some community policing advocates promote the use of local oversight boards and the placing of officers in everyday interaction with residents. The hope with such arrangements is not just to allow residents to coproduce their own security and make officers more accountable but to establish more civil social bonds between police and locals as well. Community policing models, moreover, replace the traditional focus on arrest statistics with that of addressing community problems.[64] Communitarian police departments are to be not autonomous and insulated but integrated with a range of institutions directed toward improving local quality of life.

Community policing methods strongly characterized the police force tasked with providing security to New York City's public housing projects from the 1950s until the 1970s. As historian Fritz Umbach has outlined, the fairly successful implementation of community policing permitted the city's projects to avoid suffering the same blight and decline of communality that characterized project housing in Chicago and St. Louis. Indeed, Umbach depicted housing authority policing as helping residents of New York City projects sustain community life, despite the drab impersonality of their modernist high-rise urban environment. The housing authority's police department was highly decentralized: Officers were housed in on-site "record rooms"—typically commandeered apartments. Officers were, at the same time, accountable to civilian project managers and, hence, to residents. Most importantly, they walked the beat through the residential spaces of the projects, where they would chat with residents, reprimand misbehaving children (or report them to their parents), and enforce housing authority rules; they did not simply make arrests. Finally, the police force was partly staffed by residents and reflected their demographics. Indeed, the force was 60 percent black and Latino in 1975. The success of the housing authority's community-oriented approach to policing is clear not only from below-average crime and victimhood rates, when compared with New York City as a whole, but also from the fact that local antipathy toward the NYPD did not extend to housing authority officers. In fact, incidents of brutality perpetrated by housing police were relatively rare, and residents held rent strikes to agitate for *greater* police presence, not less. [65]

Like many other potentially more communitarian alternatives discussed so far, implementing effective and just community policing systems face significant barriers. They are likely to fail if not part of a multipronged strategy simultaneously focused on encouraging social ties as well as greater economic and political community. At the same time, public safety could more often be enacted outside official policing activities. Jane Jacobs, for instance, described residents of her Greenwich Village neighborhood intervening when it appeared that a man was attempting to abduct a young girl. Producing and reconstructing neighborhoods that better encourage street life and public acquaintanceship enables residents to provide for their own collective safety without necessarily involving the criminal justice industry.[66]

Although the comparison may seem odd, CHLs as social technologies are similar to companion robots and "synthetic humans," which run the gamut from robotic seals provided to nursing home residents to interactive sex robots and noninteractive Real Dolls. However reasonable or benevolent it might seem to provide lonely seniors with a sense of companionship through an artificial pet, or a virtual lover to those without a romantic partner, such technical fixes further entrench social belonging and connection as a personal responsibility achieved through individual consumer purchases. Nursing and senior home residents are typically lonely because of contemporary education and employment systems, which lead many adults to live far away from their parents and often leave them with little time to either care for or visit elderly relatives. Inflated cultural expectations for privacy and self-reliance as well as the wider erosion of conflict negotiation skills render some retirees hesitant to move in with their children. In cities with poor transit systems and a sprawling urban landscape, an inability to drive leaves the elderly increasingly unable to seek out public sociality. The chronically lonely may turn to robotic companions because they lack the interpersonal skills and mobilities to succeed in a networked individualist society.

Such devices, in any case, are unlikely to lay the groundwork or otherwise eventually lead users toward thick community. Consider "Davecat," a man who maintains relationships with two Real Dolls for whom he has invented background stories and personalities.[67] Even though Davecat insists that his relationships with his dolls are authentic, in the end they remain fictive solipsisms. He does not commune with his partners but consumes them like any other gadget or market good. He alone controls their subjectivity, which frees him from the need to contend with the difficulties of human relationships. Indeed, Davecat has justified his lifestyle with the observation that "a synthetic will never lie to you, cheat on you, criticize you, or be otherwise disagreeable."[68] However convenient this arrangement may be for Davecat, it signifies and reinforces his inability to realize relationships with nonsynthetic humans. Surrounding himself with riskless, wholly agreeable, and fully controllable relationship partners, his relational skills are liable to further atrophy, and human others are increasingly likely to be seen as failing to measure up to what his gadgets offer.

Although Davecat is a more extreme example, and social robots remain merely a possibility on an uncertain horizon, his experience is suggestive of what might come to pass if companion and sex robots become more common. Some observers already argue that romantic relationships with synthetic humans are inevitable.[69] As anyone who recognizes that technological development is not autonomous will know, few things are actually inevitable. Companion robots and other "virtual other" technologies will only come to shape people's relational capacities if citizens fail to demand adequate regulation.

Communities of Technological Repair and Maintenance

Companion robots and handguns seem like extreme cases. What about consumer gadgets more generally? Digital devices, such as those produced by Apple, are frequently glued shut or closed with esoteric screws. Even when users can gain access to the internal components, their design may render them difficult to service. One example is how simply replacing a bad monitor connection on a MacBook Pro requires carefully heat softening the surrounding adhesive and using a suction cup to remove the glass screen, hopefully without breaking it in the process. Parts may be unavailable or so expensive as to discourage repair. The combination of such barriers leads most users to discard old and faulty electronics or have them shipped back to the manufacturer for fixing. This practice, in turn, discourages the existence of nearby repairpersons and their businesses, at least in more affluent nations, to the detriment of economic community and local social connections.

Working to oppose the social and environmental effects of planned obsolescence and wired-shut artifacts are those involved in the Fixer movement. Fixers meet throughout North America, Australia, and Europe at "repair cafes" staffed by volunteer specialists.[70] Though doing little to repair the damage to economic community, they nonetheless seek to encourage social community through the sharing of expertise, tools, and volunteer labor.

Difficult-to-service technologies also affect the development of gearhead and do-it-yourself (DIY) clubs, institutions whose role in local community should not be overlooked. Indeed, part of the "soulcraft" that philosopher Matthew Crawford described as inherent in working on classic automobiles and motorcycles stems from the fact that such activities typically take place within communities of practice.[71] Speed shops and similar locales can turn into third places, connecting adults and providing opportunities for youngsters to develop their interest in mechanical things. Devices that are needlessly complex, feebly constructed, or sealed shut with rivets and glue, however, ill afford tinkering by the average citizen, who in turn is unlikely to look for others to tinker alongside them. The need for expensive or need-lessly esoteric tools puts repair and modification out of reach for ordinary people. Every artifact thrown away or shipped off to a distant repair center signals the loss of a potentially more communitarian social or economic relationship.

Similarly, in less affluent nations, the well-intentioned dispersal of cutting-edge technologies to urban slums and rural villages—like nanopore water filters or genetically-modified seed—can undermine community. Such technologies encourage dependence on large-scale distant networks of expertise and manufacture from which the surrounding community and its resources are excluded. In contrast, "appropriate" or "intermediate" technologies are designed to be maintainable by the communities in which they are located.[72] Rather than enforce dependence on far away firms, tech-nologies like hand-operated water well pumps are flexible to a wider range of local repair practices. When combined with hand-operated drilling rigs, their construction is itself an act of community. Moreover, such pumps typically function only to the extent that they exist within communities cohesive enough to organize their maintenance—that is, insofar as locals can ensure that bolts are tightened, levers fixed, concrete aprons cleaned, and that children do not jam them with rocks.[73] Therefore, they are tech-nologies that can serve as an impetus to greater community rather than increased dependence on distant networks.

A range of techniques and devices shape citizens' prospects for realizing thicker communities. First, technologies affect people's ability to develop the necessary skills for political community by either centralizing or diffus-ing social life. Personal technologies and distributed heating systems priva-tize the domestic commons, obviating the need to resolve interpersonal conflicts. Television and companion robots provide users parasocial forms of intimacy that demand little obligation, displacing time and energy for building local social networks. Second, technologies can contribute to citi-zens' expectations for privacy, independence, and individuality. The crying-it-out method and the provision of solitary bedrooms each help reproduce cultural valuations of privacy and the appearance of self-reliance, elements of thin communitarian moral orders. The restriction of youth autonomy to certain retail, transportation, and communication systems enculturates the belief that freedom is achieved through individualizing and impersonal networks. Third, technologies like companion robots, concealed-carry handguns, A/C, and ICTs provide private retreats from more communi-tarian forms of sociability and safety, which entail practices of talk and political community, among others. Technological retreats frame goods like

social support, safety, cooling, and interaction as personal responsibilities
fulfilled by purchasing the right technology rather than as a public good.
Insofar as these technologies promise an alluring individual escape from
risk, effort, or dependence on others, they seduce users toward cocooning. Finally, technologies that are excessively complex or difficult to repair
and modify limit the development of third places, local social connections,
and economic community by decreasing the ability of ordinary citizens to
develop and maintain gearhead or DIY groups.

Surrounded by domestic and personal technologies more compatible
with cocooning than communing, many children's lives are essentially
apprenticeships in networked individualism. Even motivated communitarians are likely to struggle with the barriers posed by phones designed more
for shoring up networks than developing local solidarities and the allur-
ingly riskless substitutions for human connection offered by self-checkout
kiosks and virtual companions. Most of all, the proliferation of ever more
technological activities and digital distractions means that members of
technological societies face an increasing range of opportunities, though
mainly with respect to how to spend their leisure time. If research showing how an overabundance of choice leads to less thoughtful choosing
and more decision regret is correct, not only are people pushed to divide
their leisure time across an ever greater number of possible diversions, but
they are less likely to choose the kinds of activities they might actually
want.[74] As a result, many citizens are either stretched too thin to squeeze
community engagement into their schedules or have given the possibility
little thought, being overwhelmed by the myriad ways in which they could
spend their evenings and weekends.

At the same time, the technologies discussed above each present a kind
of social dilemma regarding community. They undermine community
as a publicly provided social good in the same way that jumping over a
subway turnstile undermines public transportation. While choosing backyard porches, concealed-carry handguns, or the cry-it-out method over
their more communitarian alternatives appears insignificant at the level
of the individual, the aggregate effect is an overall thinning of community. It is unlikely that citizens employing these technologies have foreseen,
intended, or perhaps even desired such a result; nevertheless, the combined
effect of their technological practices is the undermining of the conditions
of possibility for thick community. At the same time, it is hard to blame
people for doing what is within their grasp, even if it contributes to advancing networked individualism, given the broader forces that make realizing
thick community difficult.

In any case, technological society could better support the development of communitarian beings. Such societies would enable parents to be able to choose cosleeping arrangements more often, encourage more hearthlike heating and cooling devices, and discourage isolating patterns of leisure. Community policing could be pursued as an alternative to concealed-carry licensing. Neighborhoods could better support public safety and sociability by including public and semipublic places to cool off on hot days, namely front porches. Creating more opportunities for citizens to practice thick community would obviate at least some of the felt need to use ICTs; such neighborhoods would better satiate residents' needs for social intimacy. Finally, alternative playgrounds similar to the Land could provide children with more outdoor play opportunities that are semisupervised but otherwise self-organized. Such spaces would meet parents halfway on their "stranger danger" fears and urge to bubble wrap their children. They would provide kids with more exciting forms of play as well as the chance to learn to negotiate conflicts and be "bigger brothers and sisters" to their younger companions. Such changes would put citizens into more frequent interaction, support bonding, encourage the development of the skills of political community and talking, help center local community life, and provide the basis for a more communitarian moral order.

***

Philosopher of technology Albert Borgmann tells his ethics students that "the most important decision you can make as a young couple is whether you are going to get a television and, if you do, where you are going to put it."[75] Though this claim might be slightly hyperbolic, Borgmann's point is to get students to think about how using certain technologies influences the character of their lives. Purchasing a television and placing it in the center of one's den signals that the room is more a space for viewing than for conversation. A television in the bedroom can end up becoming a distracting barrier to arguably more desirable intimate activities. Ownership of a TV and its placement in the home is never neutral: it reflects particular ideas about the good life and reinforces certain patterns of daily living.

Through an analysis of the enablements and constraints posed by different techniques, artifacts, infrastructures, and organizations in relation to the different dimensions of communality, I have illustrated their nonneutrality with respect to the practice of thick community. Urban form, infrastructures, organizations, techniques, and gadgets do not simply provide users access to needed and desired goods and services but suggest and support only certain communal forms. Suburbia erects all sorts of physical barriers to talking and dense webs of social ties as well as effectively excises

Table 6.1

Interconnections between Artifacts and Techniques and Dimensions of Community

Dimension of Community	Thin/Barriers	Thick/Enablers
Social ties	Telephone, ICTs, "distributed" heating systems, apps like Cloak and Split, A/C, and private pools	Woodstoves, kotatsus, and other centralized heating technologies; parenting by neighbors
Exchange/ support	Concealed-carry firearms, companion robots	Self-organized play spaces
Talk	Mobile digital devices and other ICTs, self-service kiosks, conversations while watching TV or online video	Public cooling technologies like front porches, plazas, and fountains; communal and family dinner table
Symbolic/ psychological	Television, concealed-carry firearms	Public cooling technologies; collective viewing of sports/film by friends and neighbors
Economic	Hard-to-repair or unrepairable artifacts given resources of community	Intermediate technology, serviceable technologies
Political	Personal electronics, "distributed" heating systems, large affordances for privacy	Centralized heating, self-organized play spaces, community policing
Moral	Concealed-carry firearms, companion robots, cry-it-out method, private bedrooms, and child mobility achieved via networks	Community policing, child mobility achieved through community embeddedness

economic and political community out of neighborhoods and places it in distant shopping and municipal centers. Infrastructural and organizational technologies, whether an energy system or a recreation center, can frame users as either individual nodes on a network or embedded in a community. Too often they fail to coincide with the boundaries of geographic thick communities in scale, governance, and the provision of economic benefit. Gadgets and techniques can inculcate high expectations for privacy, personal convenience, and independence. A concealed-carry handgun helps users sustain the fantasy that they alone can ensure the safety of their families and facilitates a withdrawal from public, collective systems for dealing with crime. The cry-it-out method, contemporary schooling, and household spaces that support diffusion and isolation help produce adults increasingly steeped in myths of self-reliance and independence as well as decreasingly able to tolerate copresence and conflict. This, in turn, helps construct and support a moral foundation from which forms of private living, such as through suburban living or solo apartment dwelling, appear to be natural and eminently desirable.

Such technologies are not the sole contributors, of course, to the thinning of community in technological societies. Mass media advertising generally promotes a culture of buying and having over one premised on being and relating. Community becomes an afterthought when people are pushed to think of themselves as consumers first and foremost, or whenever the demands of consumer society make it difficult for citizens to carve out the necessary time in their schedules. The fact that many people can acquire several credit cards with dangerously high credit limits, alongside other cultural drivers, has resulted in an American citizenry drowning in consumer debt and thereby needing to work longer hours just to stay afloat—especially where wages are stagnant or declining. The fact that developers keep building ever larger homes and that banks continue to offer easy mortgages for them, at least in North America, means that homeowners spend more of their days dreaming up ways to fill their domestic environs with things, time that can no longer be spent on local forms of socializing. Such factors combine to produce not only everyday lives ever more defined by individual financial precarity and longer work hours but a culture increasingly preoccupied with stuff.

Specialized job markets and educational opportunities pull people out of their communities to pursue a livelihood and a career. These patterns are exacerbated to the extent that financial capital and large corporations are increasingly mobile. Feeling little obligation to the communities in which they are located, many firm owners are quick to uproot workers or move factories and offices to take advantage of tax and subsidy deals elsewhere. Contemporary demands for worker mobility destabilizes local social belonging. Indeed, citizens who frequently change residences have fewer ties with neighbors and a lower sense of community, which in turn contributes to an individualistic ethos that promotes future mobility and drives an attraction to chain and big-box stores.[76] Greater economic instability, more work hours, and tighter budgets mean less time and psychic energy left over to devote to civic involvement.

Things could be different. Technological societies could better support thick community and the development of citizens as communitarian beings. I have mentioned potentially more compatible technologies throughout the last three chapters: walkable mixed-use neighborhoods, locally governed community infrastructures, third places, neighborhood-level recreation, nondistributed heating systems, and community policing, among others. As promising as these alternative technologies might be, their promise is insufficient to ensure broader deployment and use. A constellation of factors nudge citizens' toward networked individualism, despite any desires they might have to live otherwise and, indeed, often prior to them even being able to reflect on what a desirable way of life would look like. These factors have deep roots in the political, cultural, and economic structure of technological societies. The aim of the next section of this book is to analyze how citizens might begin to strategize around the myriad social barriers to more communitarian technologies.

7 Lessening the Obduracy of Networked Urbanity

Achieving societies that enable and encourage thick community requires making everyday technologies more compatible with it. Today's built environments too often discourage neighboring, walking, and dense networks of local social bonds; media and communication devices tend to support cocooning better than communing; infrastructures organize citizens as unencumbered selves; and dominant child-rearing techniques help enculturate an overdeveloped attraction to feelings of self-reliance and privacy. Instituting more communitarian technologies, however, is no simple task. Significant barriers stand in the way. Communitarians will need to direct much of their efforts to attacking the underlying social, economic, cultural, and political drivers of the technological thinning of community life.

Viewing technological change as sociopolitical, not just technical, remains uncommon in popular debate and discourse. Many people embrace a *fairytale understanding of innovation*, believing that the purportedly "best" technologies emerge from the ostensibly "objective" processes of industrial design and market-led investment and diffusion. This belief often rears its head in the widespread claim that governments need to take a "hands-off" approach to innovation and not interfere in the "free market," even though economists long ago revealed the underlying Panglossian logic of such ideas.[1] The following chapters aim to undermine such fairytale understandings of innovation, namely with respect to communitarian technologies.

The stability and direction of technological development is guided by entrenched, pervasive, and often significantly flawed patterns of thought; systems of regulation, tax, and subsidy; already established and "sunk" infrastructures; as well as the interests of powerful sociopolitical actors. In the language of STS, the sociotechnical status quo is often *obdurate*, or difficult to change, and certain technological trajectories gain more *momentum* than others, often leading to rapid sociotechnical changes but in fairly constrained ways.[2] For instance, the ongoing and rapid innovation of frequently toxic petroleum-based chemical products contrasts the slow development of more environmentally benign and nontoxic alternatives to petroleum-based products through green chemistry. This has not been simply because green chemical processes, including the use of water or supercritical carbon dioxide as a solvent, lack promise or cannot perform adequately. Rather, the slowness of green chemical innovation partly stems from the existence of already established petroleum-processing and refining infrastructures as well as the lack of attention paid toward toxicology and chemistry ethics in university curricula and professional certification examinations.[3] Green chemistry is hindered more by the sociopolitical challenges in altering ways of thinking among chemists, funding organizations, educators, and business managers and the costs of retooling industrial civilization as we know it than the technical challenges entailed in getting green chemistry to work. The attempt to realize more communitarian technologies is no different.

In the next chapters, I provide a *reconstructivist* look at community within technological societies.[4] I discuss various barriers to more thickly communitarian technologies: cognitive, technical, cultural, sociopolitical, and economic. Such technologies are hindered by ways of (non)thinking, material or physical constraints, cultural preferences, the interests and dysfunctions of powerful political actors and institutions, as well as wrongheaded policies. Wherever possible, I suggest ways of mitigating or lessening the effects of these barriers. Although I fall short of providing an unequivocal roadmap to thicker communities, those seeking change in their own communities, cities, and nations should be able to extend my analysis and develop strategies better matched to the particularities of where they live.

Reconstructivism has yet to be applied to most of the sociotechnical problems affecting humanity. This line of inquiry within the "engaged program" of STS remains underdeveloped.[5] Those researching sociotechnical change typically catalog problems rather than imagine solutions. This tendency afflicts community studies more generally. Many otherwise excellent works characterizing communal decline within technological societies could do more exploration of how such a decline could be opposed or partly reversed, often providing a few relatively weak prescriptions in their concluding pages.[6] In contrast, chapters 7 through 9 in this volume aim to directly inform and assist citizens who wish their lives were more communitarian, beginning with urban form.[7] Unless partisans for thicker communities systematically, collectively, and directly confront the sociotechnical barriers to the ways of life they desire, they should not be surprised if networked individualization and the thinning of community life continues unabated.

***

As architectural historian Howard Davis has pointed out, buildings are cultural products.[8] Only people's complete immersion in the standard practices of today and their lack of alternative reference points make the construction of sprawling suburbs, downtown high rises, and seas of parking lots appear to be as natural as the sun rising in the morning. Such urban form is the water within which an increasing number of citizens swim. Increases in the area of urban settlement and vehicle miles traveled have significantly outpaced population growth for decades. Such patterns, moreover, have spread far beyond North America: Europe has its own sprawl, especially in Ireland, Spain, and Portugal, and builders in poorer but economically growing countries like China and Brazil increasingly build gated high-rise complexes and detached homes.[9] Regardless, because this kind of urban form is a cultural product, it could be otherwise. Buildings, streets, and neighborhoods could be redesigned to support and reflect a thick communitarian culture.

Such possibilities are erased by fairytale understandings of technological development. It is difficult for some to even fathom that large-scale suburbanization and the razing of medium-density urbanity to make way for high rises and big-box stores *may not have been* the result of simply providing "the people" what they really wanted. The view that urban form results from consumer choices, of course, has a modicum of truth to it. At the end of the day, building owners and tenants choose to sign mortgages and leases. Nevertheless, it fails to recognize that most people act within a building culture whose features are largely outside their control. How much agency do ordinary citizens really have concerning how their homes and businesses are constructed, especially when the character of surrounding neighborhoods, how bankers and investors decide which styles of building merit financing, and whether architects and contractors have the requisite expertise, imagination, and regulatory support needed to create alternative designs has already been largely decided? How is it accurate to say that the majority of North Americans have "chosen" the detached suburban home when most nonsuburban housing is located in either blighted urban areas or upscale neighborhoods? Even those who ardently desire a more communitarian environment cannot be reasonably expected to risk their physical or financial security for better urban form.

As is the case for most consumer markets, the majority of people are only able to choose among the residential options that entrepreneurs and businesses, in response to governmental regulations and incentives, are willing to produce.[10] What exactly businesses produce, in turn, is only partly dependent on citizens' "actual" preferences. Managers and owners, like consumers, act in response to a multitude of incentives, motives, regulatory pressures, and cultural ideas. Consumers' desires are influenced by years of advertising and enculturation, not to mention their level of ignorance concerning or inexperience with alternatives. In the same way that a factory is not neutral with regard to what can be produced with it—one set up to assemble automobiles could only manufacture diapers with considerable effort and expense—the vast sociotechnical networks shaping building processes are not neutral with respect to the shape of resulting urban form. How do the design features of the "urban form factory" system bias it toward building urban environments unamenable to thick community? How might this bias be lessened or reversed?

Subsidies and Perverse Incentives

The development and continuation of suburban growth has been sustained via massive subsidies and other financial incentives. Pamela Blais has described in great detail how residents of dense urban areas, namely Canadian cities, are forced to subsidize suburbia through the mispricing of development fees.[11] Most services—including water, electricity, mail, road construction and maintenance, and electricity—are characterized by "economies of density." That is, providing low-density areas with infrastructure, policing, and other services is much more expensive. This is true without even considering externalities like environmental damage, road fatalities, the health costs of sedentary and automobile-focused living, and wars in the Middle East.[12] It might cost twenty-two thousand dollars for a city such as Albuquerque to provide infrastructure to a suburban house on the urban fringe but only one thousand dollars for a home in a more central, higher-density location.[13] Development charges are routinely assigned, however, without taking such diseconomies into account. Impact or development fees could be set proportionally to the strain they put on municipal infrastructural services, and thus budgets.[14]

The perverse subsidization of sprawl likewise occurs through taxation and utility service charges. Standard mechanisms for assessing tax liabilities typically result in apartment dwellers paying more on average, despite the marginal per capita costs of snow removal, street cleaning, emergency services, and other municipal amenities being much higher in low-density suburbs.[15] Service charges for water or electricity, moreover, usually make little distinction between an area requiring ten feet of water main or electric conduit between neighboring homes and one that demands fifty or a hundred. Simply requiring suburbanites to pay their fair share of infrastructural costs would constitute a large step toward better enabling alternative forms of housing.

In many jurisdictions, however, there are no neighboring dense urban areas to subsidize diffuse growth, yet the taxes and fees remain low enough to attract homebuyers. How can those charges be so low when there are no higher-density areas to cross-subsidize them? For such places, future growth is continually sought in order to collect taxes and fees to pay for past infrastructural development. Urban political economists John R. Logan and Harvey L. Molotch called such projects "infrastructure traps."[16] The financing of infrastructure traps bears a strong resemblance to a Ponzi scheme, a form of investment fraud in which initial investors are paid their returns by the investments of later participants. Ponzi schemes, however, inevitably crash because returns on investment can only be sustained by an exponential growth in new investors. This inherent instability has not prevented municipalities from financing their infrastructural projects in Ponzi-like arrangements. Tax liabilities are often set in ways that prevent municipal revenues from making up the original construction cost during an infrastructure's lifetime. One analysis found that it would take seventy-nine years for the taxes collected from the residents living on a suburban street to pay for it, some fifty years longer than the expected lifespan of the road.[17] Expecting future growth to finance past infrastructure therefore is not only potentially irresponsible but also hides the true economic cost of sprawl.

Besides more accurately pricing infrastructure and development, denser and more communitarian development could be encouraged by increasing the taxes on land in comparison to building improvements via split-level taxation. The relatively low taxes commonly assigned to land incentivizes speculative land owning and low-density construction. Low land taxes make hoarding or not fully developing vacant land inexpensive. Comparatively high taxes on building improvements penalizes the higher-quality construction required by multiunit and compact forms of housing, promoting the inefficient use of land by diffuse suburban housing, surface parking lots, and big-box stores. Studies of the implementation of split-level taxation in Pennsylvania suggest that it discourages land speculation—keeping property values more manageable—and encourages denser development.[18]

Care needs to be taken, however, when enacting land value taxation so that more desirable low-density endeavors, including greenhouses, farms, and greenbelts, are not discouraged.

While taxation and price systems certainly matter a great deal, the largest contributor by far to the production of thinly communitarian urban form is the ongoing subsidization of automobility. U.S. gas taxes and tolls cover only roughly half of the 150 billion dollars that state and local governments spend annually on roads and highways, the rest coming from income, property, and sales taxes.[19] Seemingly unbeknownst to conservative commentators in the United States, who balk at the paltry public subsidy given to Amtrak, the costs of driving are socialized. The United States and countries like New Zealand are unique in this regard; European nations frequently collect taxes and fees above and beyond what it takes to build and maintain their road networks.[20]

Another, often overlooked, subsidy to automobility is the provision of "free" or underpriced parking. The costs to build and maintain parking spaces are bundled into rents, the prices of goods and services at brick-and-mortar stores, and local taxes. The construction costs of a typical apartment complex and commercial building can be increased 20 and 60 percent, respectively, by the need for parking facilities.[21] Nondrivers end up subsidizing automobile owners' parking costs, as when two dollars was added to the cost of the train fare to Pearson International Airport to compensate the airport authority for anticipated declines in parking revenues.[22] Planning scholar Donald Shoup estimated the hidden costs or subsidy of ostensibly "free" parking to be at least 127 billion dollars annually in the United States.[23] This figure might sound outrageous at first, but a single parking space costs between five thousand and a hundred thousand dollars to build, averages several hundred dollars in maintenance every year, and tends to occupy land that could be used for more productive ends. Moreover, countries such as the United States average three parking spaces for every automobile. To induce consumers to agitate for walkable neighborhoods and better transit, it would be necessary to increase gas taxes and make parking costs visible rather than hiding them in rents, taxes, and the price of goods and services.

The subsidization of suburbia does not end with cars: Numerous consumer-level incentives exist within the real estate industry. The ability of Americans to deduct mortgage interest from their federal taxes subsidizes exurban McMansions. Indeed, the benefits of this tax deduction accrue mainly to white suburbanites in the upper income quintile.[24] Given that the deduction only applies to citizens owning a home expensive enough for their mortgage interest to exceed the standard IRS deduction, this outcome is not at all surprising. The policy might be fruitfully reformed by turning it into a tax credit rather than a deduction, like those currently offered for efficient appliances and solar panels, and directing it toward housing in denser, more walkable areas.

Other countries have similar sprawl-friendly policies. In Canada, first-time homebuyers are offered a break on the GST (goods and services tax) of their purchase. Because this applies only to the purchase of newly constructed houses and not repairs or modification, this program discourages the buying of older homes needing renovation, which also tend to be located in denser areas.[25] Homeownership could obviously be encouraged without so strongly funneling new homebuyers into the suburbs.

More generally, the determination of potential homebuyers' capacity to service a home mortgage is itself biased toward suburban housing. Suburban living looks more affordable on paper than it really is. Neither increased heating and cooling expenses nor commuting costs are included in the relevant calculations. The different realities of suburban and urban living are not factored into the process. The energy savings of a more compact home combined with not having to own two cars, if any, should translate into people looking to live in mixed-use, transit-oriented locales being able to take out a much larger mortgage than suburbanites. Indeed, one Canadian mortgage broker estimated that the total lifetime expenses incurred by a couple purchasing a $500,000 suburban house are roughly the same as a couple buying a $720,000 urban home, if the extra costs of commuting and owning two vehicles rather than one (though not increased energy use) are factored in.[26] Accounting for such expenses when approving mortgage loans would encourage more citizens to look outside the suburbs when buying a home.

Even with these changes, urban properties may remain beyond the financial reach of the modal citizen. Partly driving their unaffordability is the extent to which central city apartments and condos are held by absentee owners, those wealthy enough to afford multiple homes or foreign investors looking for a profitable, nonconfiscatable investment in overheated housing markets. Available housing stock for residents is further depleted, and hence prices inflated, in desirable areas like San Francisco as entrepreneurs buy up and convert apartments into Airbnb rentals. Indeed, in many metropolises, large portions of the urban housing stock remain vacant most of the year. For instance, more than twenty-two thousand units in Vancouver and up to 30 percent of the apartments in the Upper East Side of New York are mostly unoccupied.[27] Given that these vacancies are typically large luxury apartments, they take up space that could be made into several times the number of affordable housing units. These practices decrease the available supply of housing, raising prices for everyone, as more building space is dedicated to luxury units that house fewer tenants and sometimes no one at all. One way to limit this phenomenon would be to increase property taxes on absentee owners—a move currently being attempted by the Israeli Knesset—or give owner-occupiers a tax break.[28] A more radical solution would be for local governments to seize vacant luxury apartments via eminent domain, an approach argued by some legal scholars to meet the "public purpose" criterion for such action: it corrects undesirable market distortions and could be used to create more affordable housing.[29]

Further driving the real estate industry toward sprawling urban development are current market-led pricing systems for urban and rural land. In Europe sprawl is enabled by the relatively low prices of agricultural and other rural lands compared to urban lots.[30] This is unsurprising given that land markets generally do not factor in the long-term environmental and social costs of paving over quality farmland and other green spaces. Indeed, research suggests that an increasing level of sprawl has been one of the biggest factors driving ecosystem damage and desertification in Spain.[31] Spanish suburbs probably would not have grown so precipitously if land prices had reflected the value of natural and rural areas as hedges against desertification. Moreover, developing greenfield sites is typically less expensive than urban infill projects, where some predevelopment remediation work may be necessary. Regionally coordinating development fees so that greenfield construction subsidizes the remediation of urban brownfield sites could slow the growth of sprawl by decreasing the costs of infill densification.

Such changes would have the added benefit of reversing the pervasive regulatory pattern wherein governments use citizens' tax dollars to make sprawling development even more alluring to developers. High upfront capital costs and long project times make the construction of high rises, shopping malls, business parks, and planned neighborhoods a risky venture. Hence, it is not surprising that such projects proceed more quickly when the financial risks are socialized (i.e., covered by taxpayers). Although federal, state, and municipal government subsidies for urban projects often have the stated goal of eliminating blight or creating more affordable housing, the beneficiaries frequently end up being affluent suburbanites rather than the poor.[32] For instance, Urban Development Action Grants offered by the U.S. Department of Housing and Urban Development have provided millions of dollars for the construction of office buildings, ski resorts, shopping centers, posh hotels, and several corporate headquarters. Municipalities, moreover, compete with each other in the attempt to woo large private developers with tax-exempt economic development bonds and "special tax districts," which force residents to subsidize large building projects while giving them little input concerning what actually gets built. The ability of "growth coalitions" of wealthy developers, contractors, bankers, and other powerful actors to elicit subsidies from politicians and bureaucrats in municipal and state governments[33] biases urban policy making, rendering it harder to incentivize the development of affordable urban housing.

One of the financial tools most frequently used to subsidize sprawl, regional malls, and high-end condominiums is tax-increment financing (TIF). Originally developed as a means to induce housing investments in blighted areas, today TIF is regularly utilized by suburban municipalities to compete for development. TIF works by funneling all or some portion of the post-development increases in sales or property tax revenue (often tens to hundreds of millions of dollars) back to retailers and developers for a fixed period, which might be as long as several decades or until the original development loans are paid off. Already profitable suburban malls frequently get labeled as "blighted" for having too few anchor stores or not being as nice as a competitor opening up across town, resulting in public redevelopment dollars originally intended to improve urban housing being used instead to subsidize the building of an Abercrombie or Louis Vuitton store location.[34] Moreover, TIF financing often ends up supporting suburban greenfield development rather than infill or densification, sometimes developments that would have probably been built anyway.[35] Besides making suburban and regional mall development cheaper and thus more alluring, TIF subsidies divert funds away from municipal coffers that could support public services like schools and transit. Certainly, TIF could be used to subsidize infill densification, build walkable neighborhoods, and convert strip and regional malls into mixed-use developments. However, the state laws permitting its use would need to set strict conditions, otherwise TIF meant for increasing densities ends up subsidizing the expansion of regional malls or the construction of McMansions on slightly smaller lots.

Dominant Frames, Embeddedness, and Persistent Traditions

A range of other obduracy-inducing sociotechnical factors drive thinly communitarian urban development. STS scholar Anique Hommels has characterized three sources of sociotechnical obduracy: sociotechnical embeddedness, dominant frames, and persistent traditions.[36] Each of these sources of obduracy help explain the ongoing momentum of thinly communitarian urban form.

Embeddedness
Embeddedness is the most straightforward cause of obduracy: it is the extent to which status quo technologies are embedded in other social and technical systems. Technologies become difficult to alter because doing so demands expensive or difficult changes. Mixed-use, medium-density neighborhoods are challenging to build because residential construction is firmly embedded in established zoning and code systems as well as existing transportation networks. Non-sprawling designs typically require expensive applications for "variances," because their designs usually conflict with local codes. Realizing neighborhoods that more substantively encourage walking, moreover, would entail altering local transportation networks to discourage driving and encourage public transit as well as reconstructing retail networks to lessen competition coming from regional shopping centers. The opposition of those with a stake in automobility, like construction firms, certain developers, and owners of big-box stores and regional malls, makes this difficult. As a result, many new urbanist developments end up being semisuburban pockets of walkability within a sea of automobile-centric development, failing to draw residents out of their automobiles.[37]

The sheer costs and difficulty of altering already established infrastructure is another aspect of embeddedness-related obduracy. The conversion of the average suburban neighborhood into a medium-density, walkable urban area would be hampered by the tangled curvilinear streets meant to evoke the sense that one is driving through the countryside. Altering street systems to better enable pedestrian traffic is, in turn, constrained by the fact that houses and backyards stand in the way of more direct roads and sidewalks. Ground infrastructures, moreover, tend to follow these roads and would need to be relocated as well. Nevertheless, the effects of material obduracy should not be overstated: sprawling apartment and business parks, malls, and residential areas could be incrementally retrofitted.[38] Large parking lots and building setbacks provide space for mixed-use development and erecting "liner buildings" close to sidewalks and the edges of streets; cul-de-sacs could eventually be connected; and suburban arterial roads could be narrowed to allow bike paths and transit lanes.

Ensuring that aging suburban infrastructure is not simply replicated but instead retrofitted could be accomplished by not only, as mentioned above, making sure development fees and land prices encourage it but also, as discussed below, establishing the right codes and zoning laws. Municipalities with the requisite political capital might consider dividing suburban lots or taking land and parking spaces bordering the street by eminent domain and selling it to those interested in constructing compact urban form—though few are likely to be so bold. More subtle policy changes could include applying a special tax levy on properties in sprawling areas that are a top priority for partitioning or retrofitting: those located on main thoroughfares or near transit stops.

Status quo urban form is obdurate as a result of its embeddedness in financial networks as much as in material ones. A case in point is the American real estate market, which has become increasingly financialized. Wall Street firms routinely trade multimillion dollar agglomerations of real estate equity and debt through real estate investment trusts and commercial mortgage-backed securities.[39] This kind of financial trading, whether applied to real estate or wheat, only proceeds when the traded object can be commoditized. As a result of financialization, American real estate has been standardized into roughly nineteen different product types, including the detached home, the garden apartment, and the retail center. This practice drives the continued obduracy of currently dominant urban form because it limits financing options for buildings not easily placed into one of the standard types. Banks are more reluctant to finance construction that they cannot easily turn around and package into a real estate investment trust or commercial mortgage-backed security. Thus, for alternative and more communitarian urban forms to stand a chance of being significantly developed, alternative funding pathways will likely need to be created.

Dominant Frames

The obduracy of contemporary suburbia also stems from entrenched ideas and beliefs. The different ways in which various social groups view a technology are what STS scholars call *dominant frames*. They are the product of differences in worldviews, styles of problem formulation, cognitive heuristics, and political values. These divergences in thinking and imagination are often difficult to reconcile, leading to stalemate and, hence, a continuation of the status quo. Conflict over which dominant frame ought to define urban form has been perennial, led by urbanists like Jacobs and Mumford as well as more recent new urbanists. The status quo, nevertheless, remains that of the "American dream," which presumes a "natural progression" of life toward the goal of the detached single-family home in the suburbs. This dominant frame guides the thinking of housing producers, motivating them to be quite conservative.[40] The American dream is incommensurable with the dominant frames of activists and planners who view ideal urban form as entailing convivial street life or the sustainable usage of scarce resources and land. This incommensurability often means that the suburban status quo remains untouched when political battles over land use patterns emerge, a result that is no doubt helped by the financial resources and privileged position of large development, construction, and retail firms in local politics. The Greater Toronto Home Builders' Association, for instance, has actively lobbied against attempts to direct residential growth away from sprawl.[41]

The obduracy-enhancing function of dominant frames is exacerbated by embeddedness. It is unlikely that building new urbanist or similarly compact urban form will enhance the feasibility of more communitarian neighborhood activity without simultaneously addressing the larger sociotechnical networks that suburbia, regional malls, and urban renewal high rises depend on. Dismantling these networks, in turn, will demand effectively opposing the dominant frames that undergird them. Consider the limited access highways that facilitate a networked individualist practice of urban mobility, fragment neighborhoods, make a suburban retreat from the city easier, and serve as barriers to walking and biking. Several cities have successfully torn down freeways and replaced them with traffic-calmed boulevards, often being able to fit bike paths, trolley lines, and additional housing in the former footprint of the highway and its on-ramps.[42] In these cases, the main barrier to removal was not financial: many were in need of rebuilding or were damaged by earthquakes, and the cost of demolition and construction of a boulevard was less than that of rebuilding. Rather, the pervasive belief that removing highways inevitably leads to nightmarish traffic congestion, which often turns out to be mistaken, typically lies at the heart of opposition.[43] Highways are seen as unequivocally "solving" the problem of traffic. The association of highways with broader aspirations and imaginaries of the autocentric good life no doubt further entrenches such assumptions, producing a dominant frame incommensurable with more communitarian visions.

Persistent Traditions

The obduracy of contemporary urban form, however, is a product not simply of ideas and the inertia of technological systems but also of a range of too-rarely-questioned commitments and practices. While dominant frames are the competing conceptualizations mobilized by rival groups over specific technologies, buildings, or systems, persistent traditions are the practices and thought patterns that "transcend local contexts and group interactions."[44] They are the more typically taken-for-granted values, understandings, and habits that tend to carry on without significant opposition or challenge. They are dominant frames that have become naturalized.

Persistent traditions are visible in the pervasive myths that sustain status quo building patterns. High rises, which often amount to vertical suburbs in terms of their communitarian compatibility, are often aggressively incentivized by municipalities and states working under the belief that high rises drive downtown economic development and solve shortages of affordable housing through increased densities. The idea that high rises consistently produce higher densities at lower costs than medium-density buildings may be just an alluring myth, exceptional cases like Hong Kong notwithstanding. For instance, despite being built out of four- to eight-story buildings, the neighborhood of Eixample in Barcelona, Spain, houses twenty-thousand more people per square mile than Manhattan. The Le Plateau-Mont-Royal borough of Montreal is one of the most densely populated neighborhoods in Canada (more than 30,000 per square mile), even though three-story walk-ups make up a large portion of its housing stock.

Such counterexamples only appear surprising because contemporary building traditions carry the naturalized assumption that high rises are the surest route to higher and more affordable densities. The factors that limit the influence of tall buildings on density and housing costs are fairly obvious on reflection. High rises are typically spaced farther apart, being surrounded by large expanses of landscaping, parking lots, and/or large boulevards. Moreover, they often include features like elevators, extra emergency stairwells, underground parking garages, and complex HVAC systems that take up valuable floor space, and they are costly to build and maintain. Indeed, according to the Canadian Mortgage and Housing Corporation, walk-up apartments cost less to build per square foot and have smaller energy footprints than high-rise apartments. Indeed, they typically have between one-half to two-thirds the overall development and maintenance costs.[45] Insofar as high rises do not actually increase densities, or not sufficiently to offset increased costs, they exacerbate the problem of urban unaffordability. One wonders if their main purpose is really to signal a city's ostensible "progress" toward modernity or to increase the revenues of local development firms.

Building Codes, Parking Requirements, and Zoning

Building codes and zoning practices help make contemporary urban form obdurate through both persistent traditions and embeddedness. Expectations for copious amounts of free parking are entrenched in parking minimums within municipal codebooks.[46] Such minimums are strictly enforced, despite being incredibly costly in terms of promoting sprawl and rarely determined with much empirical rigor. Planning departments will often demand two spaces per apartment or ten for every thousand square feet of commercial space, regardless of the presence of walkable streets or public transit. This means that developers must dedicate 360 square feet of parking for every condo and 1,800 square feet for every 1,000 square feet of commercial building. In the latter case, these policies usually guarantee that more land area is dedicated to parking than occupied space in new commercial construction. Moreover, parking codes can discourage the renovation of older buildings, for they begin to apply to older buildings whenever owners attempt to renovate them for different uses, often making it necessary to demolish part of the structure.

The obduracy-inducing effects of the persistent tradition of "free" parking obviously could be mitigated by making requirements more flexible. Builders might be encouraged to contribute a fixed sum to improved transit infrastructure in lieu of having to build an expensive above- or below-ground parking structure.[47] Given that a single space in a parking structure costs anywhere from fifteen to a hundred thousand dollars, this would result in every large development project potentially providing a multi-million dollar injection into local transit budgets. Developers might be enrolled by ensuring that contribution requirements do not totally cancel out the extra revenue they are likely to enjoy by being permitted to develop a larger amount of saleable and rentable floor area. Alternatively, developers could establish a fund that supplies building residents with free or subsidized transit passes in place of on-site parking. This would serve much the same function as an in-lieu contribution while more directly incentivizing transit use among residents.

Planning scholar Donald Shoup, furthermore, has recommended organizing existing public parking via business or neighborhood improvement districts, as has been done in East Pasadena, California.[48] Neighborhoods could raise revenue by setting local parking charges in ways that reflect supply and demand, using the collected revenue for desired local projects or even as a source for financing renovation and new construction projects. By being more directly connected to local benefits, such districts would push local residents to care more about parking infractions. Rather than being seen as a source of frustration or a form of extortion by distant municipal governments, parking can be framed as a local common-pool resource. To the extent that neighborhood improvement districts are democratically governed, this strategy could serve a dual communitarian purpose: discourage auto-driven sprawl and support greater political and economic community.

Parking regulations, however, may be slow to change. Free or inexpensive parking often gets treated as if it were a natural right. For many urban policy makers, anything but a mandatory one to two parking spots per housing unit is nearly unimaginable. Developers might be able to deftly sidestep obdurate parking requirements, however, by seeking permission to fulfill them "virtually." This was done in the Vauban neighborhood in Freiburg, Germany, where citizens established car-free neighborhoods: residents pledged not to own an automobile and to pay 3,700 euros to keep approximately two hundred square feet—a "virtual" parking spot—free of development.[49] Those desiring cars, on the other hand, had to pay roughly 17,000 euros for a space in a local garage. Apart from freeing developers from having to pave over every available square inch of land, this forced drivers to bear more of the social costs of car ownership. Fulfilling parking requirements virtually, moreover, sets aside land that can be used for public leisure. In the case of Vauban, virtual parking spaces have been used for community barbeques and pickup soccer. Having become valued public amenities, any future effort to turn these virtual spaces into actual parking lots is likely to be met with strong local resistance.

Minimum parking requirements are just one component of a persistent system of zoning and building codes that help ensure suburban building patterns. Zoning and codes enhance the momentum of particular kinds of urban form by freeing them from political scrutiny: certain buildings and neighborhoods get rubber stamped, but all others must apply for variances. Sprawl is all but assured when codes in many municipalities dictate that buildings must be set between forty and one hundred feet away from the street, lots must be at least seventy feet wide, and the floor-to-area ratio—that is, the ratio of building square footage to the land area of the lot—must be much less than one.[50] Municipal codes for the town of Framingham, Massachusetts, for instance, limit floor-to-area ratios to 0.32 in most areas, meaning that the square footage of any building must be less than thirty two percent of the piece of land it sits on; in certain residential areas, a detached home cannot cover more than fifteen percent of the lot.[51]

Requirements like these, put together, make proposed communitarian urban design harder to realize. Front porches are less effective in spurring social interaction when codes require that they be fifty feet from the sidewalk, if a sidewalk even exists. Large minimum lot widths and low floor-to-area ratios limit the number of possible units on every block. Moreover, the tradition of single-use zoning segregates residential from commercial and other uses, ensuring that residents live far away from everyday amenities, workplaces, as well as potential third places: grocery stores, pubs, and cafés. Detached homes with yards are often zoned next to parks, even though apartment dwellers have greater needs for parkland. Finally, the codebooks of many municipalities are long and complex, containing hundreds of zoning types and amendments. Developers save considerable time and money by using off-the-shelf status quo designs.

Simply making existing codes more flexible may help address the momentum of sprawling cities. Indeed, the developer of the pedestrian-friendly neighborhood of I'On in Mt. Pleasant, South Carolina, was forced to spend three years litigating to get out from under single-use zoning requirements.[52] Advocates of new urbanism promote alternative sets of *form-based codes*, or SmartCode, as a way to go beyond simply making medium-density, mixed-use urban spaces legal again. Such codes encourage or enforce their development, discouraging large setbacks and the dead space created by large driveways and oversized lots. They trim parking requirements, placing spaces behind buildings and in the center of blocks, in addition to permitting narrower, traffic-calmed streets and boulevards.[53] Such code and zoning changes are likely to be opposed by those who currently profit from building sprawl: already established developers, contractors, and realtors. Developing *hybrid* codes, in which form-based and conventional codes coexist in the municipality's books, represents an incremental step that may be more politically feasible in the short run.

At the same time, form-based codes sometimes still contain minimum parking requirements that undermine new urbanists' goals for medium-density walkable built environments.[54] How strict they are varies from city to city. Some lower or eliminate minimums to parking, setbacks, and lot widths, and merely permit mixed uses, while others require new urbanist design features outright.[55] Simply making new urbanist design legal, moreover, may be too weak a nudge to alter the momentum of suburbia and high-rise development. Without broader incentives and regulations that encourage experimentation or discourage conventional construction, alternative building codes may never actually get put into practice.

More communitarian building codes and zoning could be encouraged by mimicking the vast array of programs that originally drove sprawl. For instance, twentieth-century Federal Housing Administration policies mainly subsidized and insured the mortgages of detached suburban homes.[56] An equivalent contemporary program could be tied to mixed-use medium-density design. Suburban planning departments are more likely to update their codebooks if housing subsidies are poised to draw homebuyers to more compact neighboring cities. Other incentives would hit municipalities' budget lines more directly. Consider how alterations to states' energy codes for residential and commercial buildings was a precondition to receiving stimulus dollars relating to energy projects through the 2008 American Recovery and Reinvestment Act. Code changes could be made a precondition for federal urban development dollars.

Working through codes at all, however, presents certain risks to the practice of community. As I noted above, the point of codes is to insulate certain kinds of construction from political scrutiny by ensuring that a narrow range of designs pass easily through local bureaucratic processes. Maintaining this degree of inflexibility for zoning and building regulations that enforce compact development, though understandable, given the momentum of sprawl, nevertheless limits the exercise of political community. The development of form-based codes does typically begin with a *charrette*, a public and participatory urban design workshop. As planning scholar Jill Grant has asked, however, "once the design is finished and the codes set, where are the opportunities for democratic action?"[57] Indeed, most new urbanist neighborhoods are handed over to a homeowners' association (HOA) once they are developed. Such associations, as discussed in chapter 4, are frequently conservative by design, weakly democratic, and privilege the protection of the exchange-value of properties rather than their use-value. From another angle, given that any charrette is unlikely to get urban form "right" the first time, it makes little sense to render its resulting design overly difficult to change. A more substantively communitarian planning process would involve periodic reassessment charrettes within neighborhoods concerning their urban form, zonings, and codes. To avoid the conservatism of HOAs, a supermajority vote should be required every decade or so in order to retain existing rules. That is, advocates of the status quo should periodically be forced to carry the burden of proof.

Deficits in Relevant Professional Expertise

Adding to the barriers posed by the persistent tradition of sprawling or otherwise anticommunitarian design are deficits in countervailing knowledge among experts. Planners and architects could be better trained concerning the communitarian implications of urban design. In fact, several eminent planning professors and professionals have signed a statement lamenting the fact that planning students receive substantial training in public policy, geoengineering basics, and land-use law yet are provided very little guidance on how the arrangement of streets, buildings, neighborhoods, and parks affects the livability of a city.[58] Planning professionals thereby are left ill-prepared to evaluate master plans, zoning maps, and building codes in terms of their social implications.

Architectural education, for its part, too often treats individual buildings as abstract pieces of engineering and art rather than technologies structuring everyday sociality. Although curricula at architecture schools differ widely, evaluation criteria set out by the National Architectural Accreditation Board (NAAB) place little emphasis on how urban form shapes social relationships. Schools are to ensure that students understand the varied "social and spatial patterns that characterize different cultures,"[59] but there is little language indicating a recognition that architects help legislate the range of realizable patterns of social life through their building and neighborhood designs. Far more emphasis is placed on technical matters like acoustics, structural analysis, and environmental systems as well as the codes and financial concerns shaping the building process. Newly built high rises and suburbs alike will probably contain a relative dearth of the features that help establish social relationships through serendipitous interactions and shared routines when architecture and planning schools are not incentivized or explicitly required to teach designing for community as a professional skill.

Given that the actions of such experts have compounding effects on the prospects for community *as a public good*, it would be sensible for governments to push their accreditation boards and professional societies to require at least one course on urban design and community. Planning and architecture programs might be indirectly incentivized to include such topics if government contracts gave preference to new urbanist and similarly conceived design standards. The Department of Defense's 2012 Unified Facilities Criteria, for example, already set form-based standards for the master planning of military bases emphasizing walkability, mixed uses, and less autocentric sprawl. Such standards could be a requirement in other government construction projects. Indeed, public spending on residential, commercial, and office construction amounted to roughly 16 billion dollars in 2013 in the United States.[60] Although that represents a small but not totally insignificant 4 percent of the roughly 450 billion dollars in total construction spending, leveraging the buying power of governments in this direction—much like how the recycled paper industry was helped by the U.S. government's turn toward "greener" public procurement policies[61]— might lead planning and architecture curricula to devote more time to compact urban form.

Nevertheless, it would be a mistake to overestimate the degree of agency possessed by urban designers. As philosopher Robert Kirkman has noted, designers, including new urbanist architects, face a range of sociopolitical forces that act to limit their "efficacy."[62] No matter how well trained on the coupling of social relationships and urban form, designers would still have to contend with conservative developers, financiers, and planning offices, together with many other barriers to more communitarian urban spaces.

Affordability and Democratic Control

It is not enough, however, to alter the policies and practices that stand in the way of developing more communitarian urban spaces through mixed-use, compact design and walkability; such spaces need to be affordable. Surveys of planned mixed-use developments often find that they are significantly more expensive than surrounding neighborhoods, threatening to make them nostalgic enclaves for the upper-middle class.[63] This suggests that the standard model of using a single large private developer may be poorly suited to nonsprawling developments, at least for the time being.

There are two possible alternatives for achieving affordable communitarian urbanism as well as opening up opportunity for the practice of political community. First, the community land trust (CLT) model provides a way to keep urban land values from quickly pricing out lower-, working-, and middle-class residents. CLT land is owned by a community nonprofit that leases it to residents, who in turn own the building or unit on that land. The land is thus protected from the speculative investors, one of the most significant drivers of rapidly ballooning housing costs.[64] The CLT model has been successfully applied by the Dudley Street Neighborhood Initiative (DSNI) in Roxbury/North Dorchester, Massachusetts, and in Burlington, Vermont.[65] In Roxbury/North Dorchester, a family is likely to pay $1,300 per month for a three-bedroom unit, roughly half the going rental rate in nearby neighborhoods. Although Roxbury/North Dorchester feels more suburban than urban in some areas, the DSNI exemplifies how a CLT ownership model could help reduce housing costs and improve local control over development. The organization leads local planning and zoning processes, having the ability to hold up construction if the design is found to be untenable.

CLTs, however, face barriers of their own. Land in blighted or run-down neighborhoods is more often turned over to private developers than to community organizations. Hence, advocates face an uphill battle in convincing municipal leaders that the model is feasible. The successes of Burlington and Roxbury can be attributed to a socialist mayor in the former and a massive, highly motivated social movement in the latter. Another drawback to CLTs is that they face very real limits as the local supply of vacant or inexpensive property runs out. The DSNI has struggled to compete with entrepreneurs seeking to "flip" foreclosed homes; it is regularly underbid by them. The CLT in Burlington has run into similar barriers to expansion.[66] While many CLTs do receive some degree of governmental subsidy in the United States, including through the federal HOME Investment Partnership Program, the Department of Housing and Urban Development primarily focuses on public rental housing and voucher (i.e., rental assistance) programs. At the same time, national-level officials risk considerable political retaliation from the construction and real estate industries if they are viewed as "interfering" with private property markets, discouraging them from promoting more experimentation with CLTs. Without a significantly more favorable policy environment (subsidies, low-interest loan guarantees, and other assistance), the development of more CLTs will likely be plodding for the foreseeable future—at least in North America.

Another way around the private developer model involves institutionally supporting groups of citizens to develop their own multiunit buildings or cohousing via Baugruppen, or building cooperatives. Such initiatives are provided a large amount of public support in countries like Germany.[67] For instance, in Hamburg the Agentur für Baugemeinschaften Hamburg (Hamburg Building-Communities Agency) helps groups of citizens locate and purchase suitable land as well as performs part of the logistical work between architects and contractors traditionally supplied by developers. By cutting out the intermediating developers and real estate agents, building cooperatives in Vauban, Germany, found that they could lower costs by up to 30 percent.[68] The model could also be put to use retrofitting aging suburban sites, not just in new development. Given the observed reluctance of many private developers to implement urban designs more compatible with local social community as well as all the financial incentives pushing them to keep churning out status quo housing, lessening their privileged position in the process may be the most important first step in decreasing the obduracy of suburban and luxury high-rise construction.

Nevertheless, there are barriers to implementing building cooperatives in North America and elsewhere. First, they are unlikely to be successful in an institutional vacuum. Some equivalent to the Hamburg Building-Communities Agency needs to be available to guide cooperative members through the requisite logistical, legal, and financial processes. Probably most importantly, this organization needs to be prepared to mediate serious conflicts and lessen the force of the persistent tradition of relying on litigation to solve disputes. Indeed, a significant focus on conflict mitigation was a fundamental part of the planning process in Vauban.[69] Citizens accustomed to private living and the suburban/high-rise culture of avoidance will be ill-prepared to work cooperatively with their future neighbors. Financing also needs to be made available. If traditional banks are wary of loaning to nonbusiness entities for multiunit housing, perhaps local credit unions could be encouraged to take on that role. Given that many municipalities already forgo millions in revenue via tax breaks to traditional developers to little avail, there is little reason that local politicians could not direct some of that spending to underwriting loans to building cooperatives.

Promoting and Collectively Governing Third Places and Neighborhood Recreation Spaces

As I have mentioned, a communitarian neighborhood is more than just compatible urban form. Such spaces must also contain attractive and affordable places for leisure. A mixed-use street full of swanky wine bars and artisan delis may attract networked communities of hipsters and creative-class professionals, and their pocketbooks, but hardly supports thick community. Third places, the local bars and cafés that center local community life, have traditionally provided low-cost hangouts for those living nearby. Yet, they are increasingly discouraged by a constellation of financial disincentives. To begin, alcohol at grocery stores and other outlets is often sold at one-third to one-quarter of the pub price. This price difference is exacerbated wherever groceries are allowed to sell alcohol at or below cost (i.e., as a "loss-leader") to attract customers. Bars and taverns, moreover, have high operating costs, stemming from insurance rates, high rents, liquor licensing fees, and legal and contractual requirements that force them to purchase highly marked-up alcohol from only approved distributors. The widening gap between on- and off-premise alcohol prices encourages drinking at home. Even in the United Kingdom, where pubs have traditionally centered social life, people now drink twice as often at home as at bars, restaurants, and clubs.[70]

Several policy changes could help reverse this trend. First, setting minimum prices for a unit of alcohol, as is being considered in the UK, could lessen the gap between grocery store and pub prices.[71] Third place creation could be encouraged by reducing the tax burden on pubs and cafés that serve community needs or, as the City of Chicago offers, a break on liquor license fees. Indeed, fraternal lodges like the Eagles and VFW posts in states like Montana are able to offer drinks for between one and two dollars, even to nonmembers, largely because of their nonprofit tax-exempt status. No doubt there are details to work out regarding how much community service and how big of a tax break. Rick Muir, associate director of the Institute of Public Policy Research, has suggested offering tax relief of 50 percent for UK pubs that center local social networks; encourage the mixing of diverse social groups; offer public services like mail storage, meeting space, and rides to intoxicated patrons; or pursue and host charitable fundraising activities.[72] A similar set of requirements could be devised for cafés. Additional financial breaks, moreover, could be offered to pubs and cafés that embrace community ownership models that provide citizens the chance to practice democratic political community.

Centralizing the provision of certain goods and services at the neighborhood level could enable third places in neighborhoods otherwise not amenable to them by design. For instance, governments could lower the expenses incurred by their ailing postal services at the same time that they promote togetherness by ceasing to deliver mail to residents' doorsteps. Residents would be forced, much like those living in small towns in Montana and elsewhere, to visit the mailboxes at a local mail sorting center. Such centers could be placed next to or even within a small community center complete with a pub or café. Even those picking up their Amazon purchases may be tempted to stay and linger if they spot an acquaintance or are feeling lonely. The efficacy of this approach likely depends on getting the scale right. There may be efficiency gains in attempting to serve ever larger populations, but only at the cost of greater anonymity. At the same time, centers serving too few residents may never achieve the critical mass to become vibrant gathering places.

Places for playing outdoor games like soccer or horseshoes or simply to relax under shade trees further provide informal and inexpensive places for social congregation. Nevertheless, municipalities tend to centralize these features into large parks and recreation centers in order to minimize operation expenditures. Such facilities may be within reasonable driving distance but still far enough away to discourage everyday usage. Many suburbanites, moreover, may continue to believe that the private backyard can support all of a family's recreational needs, even though increasing rates of obesity and sedentary living in sprawl-based societies suggest that they do not. Bigger cultural forces in the forms of "stranger danger" and the organization of youth sports are also at play.

Regardless, changes to zoning regulations could ensure that all residents are within a half or quarter mile of some space for public recreation. Vacant lots could be cheaply converted. Alternatively, space might be opened up by taking control of superfluous parking spots or underutilized lawns. Regular operation and maintenance of these facilities could be partly handed over to neighborhood residents, along with funds for doing so or the ability to raise those funds through a neighborhood improvement district. Residents in the Water Hill neighborhood of Ann Arbor, Michigan, for example, pooled funds to purchase a SnowBuddy snow-clearing device, each taking their turn to clear the local sidewalks.[73] Similar arrangements could be made for maintaining neighborhood recreational space. In many cases, however, that level of collective organizing is not even necessary. Most suburbs reliably contain a few residents with riding mowers grossly oversized for their lawns. They might be persuaded to occasionally mow the local soccer or baseball field by the combination of a six-pack of beer and their neighbors' esteem, if not a credit on their local property taxes. Others might prefer a half hour spent flooding an outdoor ice rink to the equivalent time shuttling children to a megafacility across town.

Communitarian urban spaces, however, are unlikely to exist without lessening the barriers to more communitarian urban governance more generally. Although homeowners' associations and business improvement districts have merits, they tend to be weakly democratic and focused more on property values or local consumer spending than on improving the livability of neighborhoods. A proposed outdoor hockey rink is likely to be viewed as a probable insurance liability and an "eyesore" that somehow lowers local property values. Nevertheless, there is no reason that homeowners' associations and improvement districts could not be induced to become more responsive institutions of collective governance. I suspect, however, that many homeowners' aspirations continue to lie in the direction of an exurban home where ostensibly *no one can tell them what to do* for, in North America especially, the alluring myth of rugged individualism undermines many people's ability to see democratic action as a credible possibility. Residents' limitations in thinking and conflict negotiation skills may stand in the way of communitarian self-governance more than anything else. Business owners, for their part, will likely oppose expanding business improvement districts to include homeowners, much less tenants, because the power of their own votes would thereby be diluted. Still, if permitted the regulatory space and provided a means to financially support their efforts, neighbors could be enticed into more often working together to operate small-scale local amenities.

People have not really "chosen" sprawling suburban development. Not long after Levitt and Sons broke ground on Levittown and the first bulldozers razed Boston's West End, the momentum of suburbia and urban renewal became all but ensured through a constellation of subsidies and entrenched sociotechnical systems. I have outlined how the mispricing of land, utilities, and development leads to the perverse incentivizing of low-density construction. Developers of high-end condos and upscale suburban neighborhoods have been shrewd operators in getting public dollars to support their bottom lines. Other processes have simultaneously been at play. The way in which privileged actors—developers, planners, and political elites—understand urban form is often incommensurable with compact mixed-use development. Sprawl has its own physical momentum in the form of tons of concrete and asphalt. The arrangement of buildings in relation to roads and highways within suburban cities helps enforce the reproduction of the status quo. Finally, a range of persistent traditions, from entrenched parking and building codes to how architects and planners are educated and development is financed, make the current trajectory of urban development difficult to change.

In principle, however, policies and practices supporting thinly communal development patterns could be strategically attacked and incrementally altered. Development charges as well as gas and property taxes could be set to make suburbanites pay for the municipal services they receive. Forcing banks to include in mortgage calculations the hidden transportation and energy costs of suburban living would push more to consider living closer to downtown. National and state governments could encourage municipal and regional planning organizations to use, or at least permit, compact and walkable mixed-use development by tying federal funds and guarantees to codebook changes. Such code changes, moreover, might get more political traction if they were enforced gradually and flexibly, as in parking garages with first-floor retail that are designed to be easily converted into housing, rather than trying to realize ideal urban form at the outset. Changes to planning and architecture curricula, as well as more amenable regulatory environments, could ensure that care for community could figure more prominently in the skill sets and practices of professionals. Citizens could play a bigger role in determining the character of their neighborhoods. Through community land trusts, neighborhood improvement districts, building cooperatives, and/or participatory planning boards, residents could be enabled to ask more from local urban form, especially if designed to periodically place the burden of proof on the status quo. Finally, code and tax changes could make third places, in the form of a local pub, café, or soccer field, easier to realize.

Although the opportunity to intervene in the last generation of urban design and construction is already forgone, the chance to considerably alter the next generation of housing remains. According to some estimates, around 35 million residential units will be newly constructed and another 17 million will be rebuilt between 2000 and 2025 in the United States alone, representing more than 40 percent of the 2000 supply of houses.[74] At the same time, 77 percent of the 76 million American millennials, those born around the start of the twenty-first century, express a desire to live in city centers.[75]

The limited life span of built form provides opportunities for ameliorative action. High rises are frequently demolished or demand significant renovations after thirty or forty years—or even in as little as fifteen, as residents of glass-skinned condominiums in Toronto have been loath to discover.[76] Highways and roads need major renovations along these same timescales. Few of the cheap regional malls erected since the 1950s are likely to remain usable very far into the twenty-first century, and at least some of the homes will start to require serious repairs. Although massive changes to urban form are likely most feasible in countries like Japan, where the expected lifespan of most buildings is between twenty and forty years, intelligent steering of policy to more compact new construction and a significant focus on retrofits and infill could lead to a substantial shift in the building stock within a few decades.

Many of the barriers I discuss in this chapter have been known for quite some time, suggesting that constraints on learning and thinking might be the most significant obstacles. Indeed, I have already touched on some of the myths that legitimate the status quo: the belief that suburbia has not been subsidized but has been produced by a free market providing homeowners what they really want; that suburban living is more affordable; that cities can only cope with automobile traffic through limited access highways; that parking needs to be free; and that high rises uniquely or unequivocally solve deficits in urban density. Such myths sustain a vicious cycle.

They help ensure that alternatives are rarely built. Because most citizens and professionals have little experience with alternatives, such legitimating myths proceed unchallenged. Plain unfamiliarity and the perception of more communitarian urban form as a noncredible alternative figure as significant barriers in the minds of consumers.[77]

These mental barriers could be chipped away bit by bit. Suburban dwellers can only imagine life to be impossible without the privacy of a sizable yard and a four-lane highway separating home and work partly because they have never lived in any other way, so no one could know which counterexamples would awaken their imaginations. Planners are enculturated to place too much faith in the codes set by their forebearers, tending to view them as objectively or scientifically determined laws of urban form rather than as fallible, value-laden, and often arbitrary. Such views can be attacked directly; indeed, this work is partly an attempt at that. And advocates for more communitarian urban form might be most successful by simply putting their heads down and working to realize the technological and political changes they desire, waiting for stubborn and anachronistic ideas to die with the people who hold them.

Urban form is but one of several types of technology that impinge on the practice of community. Having more neighbors and amenities within walking distance will likely have no more than modest effects on the practice of community if most citizens continue to meet their daily needs through individualizing devices, impersonal institutions, and large-scale technical systems. In the next chapter, I turn to barriers faced by more communitarian organizations and infrastructures.

8 Imagining More Communitarian Infrastructures and Organizations

Nothing is natural about individuating infrastructures and organizations; they could be different. Much of what is provided by "modern" infrastructure could be realized without leaving out so many of the sevens dimensions of community. An infrastructure like indoor plumbing can no doubt feel like an unalloyed blessing when compared to lugging water from a nearby river, a burden that has traditionally fallen disproportionately on women. Nevertheless, there are ways of designing infrastructures and organizations so that they provide easier access to desirable goods and services without individuating citizens to such an extreme degree. How is the infrastructural and organizational status quo sociopolitically maintained?

Attending to this question uncovers how the sociotechnical change is not the result of technology simply evolving toward some predetermined end state but something complex, contingent, and difficult to predict. In particular, how do regulatory policies and economic incentives stand in the way of alternatives? What role is played by ideas about what constitutes "progress?" How might the range of actors permitted to supply energy or other goods be broadened? Could economic policies be redirected from supporting large chain stores and Internet shopping to supporting local business? How do entrenched cultural beliefs stymie citizens' imaginations, leading them to less often see local cooperation as a pathway to more desirable ways of living? In this chapter, I characterize these and other barriers to realizing thick community through infrastructural technologies and organizational systems as well as potential ameliorative strategies.

Toward Communitarian Infrastructures

Community Energy
The energy landscape in most technological societies is dominated by large power corporations and municipal utilities, albeit with a range of

credits and incentives meant to encourage individuals to produce or conserve energy within their private homes. A more communitarian energy system would group neighboring residents into energy improvement districts (EIDs) with their own energy microgrids. Residents would collectively own and manage local energy infrastructure, such as community-scale photovoltaics or wind power, within a small-scale grid that can be disconnected (i.e., "islanded") from the larger grid. Although energy microgrids are frequently found in large campus environments, like hospitals and universities, they are rarely present in residential neighborhoods. What legal, financial, and other barriers would need to be lessened for community energy to be more feasible?

To begin, microgrids, community-scale energy infrastructure, and EIDs currently exist in a regulatory vacuum. Formal policies in the United States vary state to state regarding the legality of microgrids as well as whether they are classified as public utilities and thus subject to the same regulations as large power plants.[1] The local utility usually holds an exclusive right to sell electricity, which effectively prohibits community energy. Moreover, there are numerous regulatory uncertainties pertaining to how microgrids should be connected to the larger electrical grid, what the expectations for buying or selling energy to the local utility should be, and which utility standards ought to apply to microgrid owners. Within this state of uncertainty, utilities often further discourage the development of decentralized energy infrastructures by requiring expensive interconnection studies as well as redundant and potentially unnecessary protective hardware.[2] Planning departments can add expensive delays, since most municipal zoning regulations and codes do not yet recognize intermediate-scale renewable energy infrastructures.[3] While EIDs have been formally established in the state of Connecticut, the relevant legislation leaves them institutionally stunted from the beginning: They are entirely voluntary; their creation and governance lies solely in the hands of local mayors and councils; and they are not authorized to be energy utilities.[4] Without a clear opening within regulatory energy regimes, few residents are likely to attempt to organize a local EID microgrid.

Even if policies that allow and encourage community energy systems were developed and clarified, there would still be the matter of getting the requisite infrastructure built and maintained, yet this barrier is not as onerous as one might think. Much of the energy infrastructure in North America and in other developed nations is already at the end of its expected lifespan.[5] Indeed, the American Society of Civil Engineers gave the United States' electrical system a D+ in 2013. Energy companies usually consider it more profitable to retain antiquated infrastructure as long as it contin-ues to function, because rebuilding distribution networks and generation systems requires a great deal of up-front capital and the negotiation of complex regulations. Hence, they are currently incentivized to retain as much of their already fully amortized infrastructures as possible.[6] Govern-ments could take advantage of this state of affairs to encourage commu-nity energy. For instance, they could enact a range of penalties on utilities who fail to upgrade their systems and use the collected revenue to subsi-dize microgrid installation. States, moreover, may find funding more easily forthcoming if they frame microgrids as New York is doing: as a means of ensuring energy resilience in the face of natural disasters, which are poised to become more prevalent and extreme with the advance of global climate change.[7]

The potential for community energy could be incrementally advanced by requiring that electrical networks in large new construction and infill projects be "microgridable." That is, they need not constitute a microgrid right off the bat but at the very minimum would contain the necessary electrical infrastructure to make later conversion relatively straightforward. Moreover, building codes could require that builders set building orienta-tions to ensure compatibility with solar panel installation and locate com-mon areas where solar and wind exposure is highest.[8] Going further, codes could require the building and locating of neighborhood-owned buildings so that they could later house community energy infrastructure. This would enable neighborhoods a great deal more flexibility than currently exists to pursue community energy as fossil fuels eventually become more expen-sive. Indeed, Eugene, Oregon, already requires that 70 percent of buildings in certain zones be oriented to permit the installation of solar panels. A shift in building codes and planning regulations could lay the groundwork for not merely household energy production but production at the scale of community as well.

Even with a suitable regulatory regime and compatible infrastructure, citizens are likely to pursue community energy only if it makes financial sense—that is, if community-level control either incurs little penalty or, even better, generates cost savings for members. There is good reason to suspect that such infrastructures could be economically viable. In Europe wholesale energy is often priced at three to five cents per kilowatt-hour and sold to customers for ten to fifteen cents, a markup partly resulting from transmission and distribution costs.[9] This suggests that considerable cost savings could be achieved by lessening the need for long-distance distribution—as long as production efficiency is not too dramatically affected. Moreover, because community and district energy systems serve small geographic areas, they can more feasibly incorporate cogeneration technologies that utilize waste heat from electricity generation to supply heating and cooling to nearby buildings. A cogeneration plant running on gas and chips made from tree trimmings and downed trees, for example, supplies much of the energy, heating, and cooling to downtown St. Paul, Minnesota.

As attractive as the potential efficiencies may be, residents nevertheless would need to be able to manage the initial capital costs in order to be willing to take the plunge. Part of this can be accomplished by charging resident fees through institutions like EIDs, or potentially via homeowners' and condominium associations. Low-interest loan options, grants, and high guaranteed *feed-in* tariffs for excess energy sold to the larger grid, furthermore, can help reduce the financial uncertainty of such projects. As well, carbon and other green taxes can discourage the growth of the traditional energy system. A mix of such policies, including a ban on electric heating in new buildings, helped Denmark grow the percentage of its domestic energy produced by small-scale cogeneration plants from 1 to 11 percent between 1990 and 2012.[10]

Creating a favorable environment for community energy, however, will probably face strong opposition from already established actors. Given budgetary constraints, subsidies for community energy are likely to mean reduced federal handouts to large-scale fossil fuel and nuclear plants, as well as potential declines in profit. The multimillion-dollar campaign waged over the last few years by Xcel Energy against efforts by the city of Boulder to municipalize energy production, and Hawaiian Electric Company's virtual moratorium on new grid connections of rooftop solar installations are suggestive of the kind of opposition likely to emerge against community energy.[11] Given the power and resources of large energy utilities, successfully implementing community energy would probably require that advocates organize regionally and nationally. On the other hand, advocates for community energy might be better off side-stepping this barrier—for the time being—by focusing first on electrical cooperatives, which already serve some 42 million people in the United States alone and are generally more democratically accountable to local residents than is the typical utility company.

Nevertheless, even if a favorable policy environment is created and the opposition of powerful actors surmounted, community energy systems face barriers stemming from culture and entrenched patterns of thought. Many citizens are unaccustomed to collaborating with neighbors for any reason at all, much less for their daily energy needs. For those habituated to obtaining the needed goods and services for everyday life by handing over dollars rather than organizing and negotiating, community-scale governance likely seems an ominous and frustrating burden. Efforts to promote community-scale energy therefore may not take off until other aspects of everyday life, such as schooling or recreation, begin preparing citizens for the practice of political community.

Retail Networks and Cooperative Business

As I argue in chapter 5, contemporary retail networks individuate consumers and centralize selling within the hands of large big-box chains, regional malls, and online retailers like Amazon. A more communitarian arrangement to retail would put buyers into more frequent, stable, and collaborative relationships with local producers and sellers. This need not mean going so far as instituting cooperative businesses but could merely require assisting the development and maintenance of local small-scale business proprietorship. Indeed, local proprietors are often public characters who perform important but often invisible community labor.[12] How do the activities and presence of large corporate retailers, current policy regimes, and bureaucratic and financial challenges stand in the way of local business?

Purchases at regional malls and big-box outlets are alluring in part due to consumers' cognitive limitations. Consumers are notoriously prone to failing to enact their broader value commitments when it comes time to buy something. Immediate financial concerns, time constraints, advertising effects, deeply entrenched cultural ideas, and a litany of cognitive biases shape consumer purchasing habits.[13] While a neoclassical economist might insist that decisions made at store registers "reveal" people's "real" preferences, it seems wrongheaded to view people's reflective preferences as somehow less authentic than their behavior when forced to decide between more principled purchasing and being able to afford that month's groceries. Although irrational at the level of abstract logic, practically speaking, there is nothing unusual about consumers lamenting the decline of local community at the same time as they shop at Amazon and outlet malls. It is incredibly challenging for consumers to meet their longer-term, reflective preferences within the circumscribed and often anxiety-provoking context of economic purchases. Part of the problem is that keeping track of their dollars is relatively easy for consumers, but they have no similar accounting system for intangible public goods like community.

Large retailers gain further invisible "subsidy" from supportive technologies. A case in point is driving, which is not merely appealing because of the lack of quality public transit alternatives but also as a result of amenities being spaced far apart. The fact that so many people own automobiles, in turn, further encourages shopping amenity placement at low densities. In societies where most citizens drive, it is in the interests of retailers and other providers of goods and services to distribute their locations as sparsely as possible. Doing so enables them to minimize staff: fewer stores means fewer store managers. Furthermore, this arrangement allows some of the distribution costs to be handed off to customers, who now drive to more distant stores for their purchases.

A rough estimate of this delegation of cost is not difficult to calculate. Taking the median distance between residents and any Wal-Mart (about 4.2 miles) as representative of how far the typical Wal-Mart shopper drives along with AAA's estimates of the costs of driving, then the total cost of a roundtrip to Wal-Mart runs between four and eight dollars. Given that customers typically spend between fifty and sixty dollars per trip, shoppers are actually paying somewhere around an extra 10 percent for the privilege of driving to a supercenter.[14] Such spending is, of course, already a sunk cost, "paid for" via consumers' increased gas expenses, maintenance bills, and insurance fees. Regional-level stores like Wal-Mart leverage the relative invisibility of such costs into a competitive advantage over local businesses, appearing to have lower prices than they actually do. Certainly, ensuring that more citizens are aware of the costs of discount regional retailers in terms of decreased economic community and increased driving could help matters. More substantive changes, however, would come from arranging tax incentives and penalties in order to force retailers to bear more of the costs produced by their distribution models.

The price advantage enjoyed by large chain-store and online retail competition is exacerbated by the extent to which these large economic actors are directly subsidized by governments. For instance, the construction of a Wal-Mart supercenter in Bridgeton, Missouri, received 7.2 million dollars of subsidy in the form of tax-increment financing. The conversion of a Mervyn's into a Target in Kenner, Louisiana, received 1.4 million dollars in public support. The scale of taxpayer-funded subsidy of big-box and online retail is staggering. Good Jobs First estimates Wal-Mart's and Amazon's total direct public subsidy in tax rebates, low-cost loans, and grants in recent decades to be 182 million and 575 million dollars, respectively.[15] Such figures do not even account for their public subsidization via the Medicaid dollars and food stamps awarded to poorly paid workers, or online firms' decades of exemption from sales taxation. Public assistance paid to low-wage workers at Wal-Mart and McDonald's amounts to 6.2 billion and 1.2

billion dollars, respectively, each year in the United States.[16] Given that Amazon earned around 74 billion dollars in revenue in 2013, continued sales tax exemptions in many states means that millions in forgone tax- payer dollars incentivizes online shopping at Amazon each year. Leveling the playing field between these large firms and local businesses would help promote local proprietorship or, at least, prevent economic concentration from getting worse.

Eliminating or redirecting some of these state- and municipal-level sub- sidies is easier said than done. Direct subsidies and tax rebates frequently exist because states and cities seek to compete with each other in the attempt to lure large corporations and tech firms to their area. In essence, this is *smokestack chasing*, a kind of economic development arms race that significantly tightens municipal and state budgets.[17] The inability of dif- ferent states and municipalities to cooperate leads them to greatly worsen their own fiscal situation out of fear that they might lose out on develop- ment. Large firms, for their part, pit governments against each other to ensure that the firms have minimal tax liabilities or even have their capital costs directly subsidized by the state. Twitter, for instance, extracted 22 million dollars in payroll tax exemptions from the city of San Francisco in exchange for staying put, and New York City's mayor infamously offered a 1 billion-dollar incentive package to the New York Stock Exchange when it threatened to move to New Jersey.[18] These incentives, however, rarely deliver. Governments persist in handing them out despite studies show- ing that they are often ineffective in spurring long-term job growth and frequently fail to produce net economic benefits. Subsidies for professional sports facilities are the most irrational: municipalities typically never recover the dollars spent on building stadiums and arenas for the billion- aire owners of sports teams.[19]

Although activists in some cities have managed to get local economic benefit employed as an evaluative metric when large-scale development projects are subsidized, mainly through "community benefit agree- ments,"[20] it seems clear that most individual state and municipal govern- ments are likely too politically weak or too optimistic of their ability to chase smokestacks to effectively counter the political and economic power of large firms. Like any arms race, the solution lies in developing bind- ing "treaties" or other regulations to limit interstate or intermunicipality competition.[21]

One such strategy for limiting smokestack chasing is to share some por- tion of increased tax revenues across metropolitan regions or states. The logic undergirding such measures is that governments will be less likely to lure firms away from their neighbors if they are inevitably forced to share the spoils. In fact, the seven counties of the Twin Cities region of Minnesota have pooled and equitably redistributed 40 percent of any increase in property tax revenue since the seventies.[22] While I could not find any studies directly measuring the effectiveness of tax-sharing in reducing corporate relocation, revenue sharing remains a promising avenue that merits more experimentation. Implementing this policy in other places and at the scale of states and regions, however, requires convincing current "winners" that they would be better off cooperating rather than competing. A similar, though currently unexplored, policy would be to require that every public dollar used to chase smokestacks be matched by one used to develop local business.

Apart from reducing the subsidization of large corporate retailers, economic community would be enhanced by addressing the bureaucratic and financial barriers faced by local proprietors. Many small business owners in the poor neighborhood of Camden, New Jersey, for example, are ineligible for financing through traditional channels and find navigating municipal and state bureaucracies to be a daunting task.[23] Although the Small Business Administration supports some nine hundred small business development centers across the United States, a large percentage of cities and towns have no one-stop location for free or inexpensive business consulting or help with filling out paperwork. Likewise, while the U.S. Treasury Department's Small Business Lending Fund invests some 4 billion dollars in community banks and community development loan funds, this is just a drop in the bucket compared to the roughly 180 billion dollars in small business loans made each year. Loans to small businesses as a whole have been in decline for years, with the consolidation of small banks into large firms and tightening lending standards likely being the most significant drivers. Credit unions, for their part, have stepped in to take up some of the slack. Their ability to make business loans is currently limited, however, at least in the United States. The Credit Union Membership Access Act of 1998 limits business loans to 12.25 percent of their assets. Economic community would be well served by adding an exemption to that limit for business loans to local small-scale firms.[24]

Cooperative business models go a step beyond small-scale local proprietorship by broadening ownership and decision-making power to a larger number of people. Although they are becoming more common, especially food co-ops, their number and reach could be enhanced through alternative financing mechanisms in addition to improved organization and collaboration. Certainly cooperatives would be more common if other governments followed Quebec's "social economy" model. The Chantier de l'economie sociale (Social Economy Network), along with the Quebecois government and other political actors, has supported the growth of a range of "solidarity financing" mechanisms that provide cooperative enterprises and similar models with access to small start-up loans, angel investments, and long-term patient capital. As a result, social economy businesses account for some 8 percent of Quebec's GDP. The Canadian province, moreover, provides a very amenable context for such programs, given its relatively high rate of unionization, progressive politics, and a long history of large agricultural and financial cooperatives (e.g., Desjardins).[25]

Developing a similar level of financial support and expertise for cooperative ventures in other places could be done by better leveraging existing large institutions. The Quebec Labour Federation was instrumental in developing early solidarity funds for the province's social economy movement. Unions in other areas of North America could be helpful in growing the number of cooperative businesses. In fact, the United Steelworkers of America has taken a keen interest in the famous Mondragon Cooperative in the Basque region of Spain.[26] Although emaciated compared with their numbers and political power several decades ago, North American unions are still very large organizations. The Service Employees International Union (SEIU) alone manages 1.8 billion dollars in assets in their Master Trust Pension Plan, more than is held by the National Cooperative Bank—the largest American financial firm dedicated to serving cooperatives. If unions invested half their pension dollars in funds for financing cooperative businesses, the economic prospects for such firms would be substantively changed within a decade. Unions could also provide a great amount of logistical support, given their ability to raise funds through dues, and they could help provide better benefit packages to cooperative workers by facilitating the collective purchasing of insurance. Finally, organizations such as the U.S. Small Business Administration could have their missions expanded to include worker, retail, producer, and consumer cooperatives.

Cooperatives often fail to get a more significant toehold in local economies because they lack sufficient levels of organization at larger scales. Although organizations like the National Cooperative Business Association both advocate for and provide business services to cooperatives, they could do more to help them better compete with large corporate firms. Consider ACE Hardware, a retailer's cooperative. Each independently owned and operated store is a coowner of the ACE brand and takes advantage of collective purchasing agreements with wholesalers. This model is too rarely extended to entities like food cooperatives. Even though food co-ops attempt to source much of their produce and other goods locally, they inevitably rely on large organic firms like Eden or Cascadian Farms for dry goods and a range of other products. Arranging themselves as a retailer's cooperative, in addition to being a consumer and/or worker cooperative, would enable them to better compete pricewise with large chain firms, namely Whole Foods or Trader Joe's, who have adopted "foodie" culture as part of their business strategies.

Advocacy against chains and franchises might be more successful if they framed their efforts in terms of promoting cooperatives rather than as a form of "Not in My Backyard!" protest. People in cities like Troy, New York, have taken the latter approach when McDonald's or other chains have come to town. Rarely in such cases is the idea of a cooperatively run burger joint advocated as an alternative to corporate development. Doing so would offer the added benefit of gaining the support of residents who might not be so morally opposed to fast food or drive-thrus. There is no reason that cooperatives should serve only the tastes of white middle-to-upper-class foodies. Consider how residents of Saranac Lake, New York, organized their own for-profit department store when a Wal-Mart Supercenter was poised to move in. Rather than suffer the loss of local proprietors, the town gained a community-owned business, wherein residents could purchase up to a maximum of ten thousand dollars in hundred-dollar shares.[27] Rather than lament the aesthetics and culture of department stores, Saranac Lake residents focused on ensuring that such shopping venues supported their town's economic community. Instead of merely winning a single battle against large corporate retailing by preventing the construction of a Wal-Mart, they have won the war, for the time being, by ensuring that future discount stores will have to compete with their own well-supported, home-grown alternative. Such successes will need more careful study by community sociologists and other scholars to learn how similar towns might realize the same outcomes.

The number of cooperatives could be grown by revising policies to reflect the lessons offered by Cleveland's Evergreen Cooperative.[28] The stability and success of the firm's laundry and greenhouse operations have had much to do with a long-term business agreement with an *anchor institution*: a nearby hospital. Services provided by entities like universities, schools, municipal governments, and hospitals cannot be easily outsourced, thus they represent a relatively stable focus for local economic activity. City governments could substantially grow the number of local businesses by implementing *import substitution* policies that require anchor institutions to meet some of their needs via long-term commitments with home-grown cooperatives.[29] Rensselaer Polytechnic Institute, for example, maintains an exclusive noncompetitive contract with Sodexo, a large multinational, for its restaurant and catering needs—much to the detriment of the local economy of Troy, New York. Rensselaer's administration seems to fail to recognize that gains in revenue resulting from contracting with Sodexo are counterbalanced by losses elsewhere: Troy's continued economic woes contribute to crime and depress its nightlife, driving away potential students, faculty, and the high-tech start-ups the university desperately wants to attract. There is little reason that the institute, in conjunction with the city and other stakeholders, could not help grow a local cooperative catering company to suit their needs or develop a similar agreement with a consortium of already existing downtown restaurants. Agreements like these leverage the needs of anchor institutions toward the development of more vibrant communal economies.

State welfare programs likewise fail to take advantage of similar synergistic possibilities. North American projects and other forms of subsidized housing are often located in *food deserts*. Grocery stores are rarely located in such neighborhoods. Residents must rely on convenience stores full of marked-up processed food or take long bus trips to corporate supermarkets. At the same time, they typically receive food stamps as well. Because residents live in food deserts, however, much of their food aid dollars ends up as revenue for Wal-Mart and other corporate grocers, not local proprietors. Policy makers could get more community development bang for their buck if food aid did not simply subsidize recipients' nutritional needs but helped foster economic community. Subsidized housing blocks could be retrofitted to make room for publicly supported retail space, and residents could be enticed into managing their own cooperative grocery. Moreover, policy makers could reimburse food stamp dollars at a higher value at the cooperative grocery than elsewhere to encourage patronage. This would not only create local jobs, sorely needed by project residents, but also help recirculate welfare dollars within the community, spurring greater economic stability and growth. Such places need not be mere stopping points for government dollars on their way to the registers of corporately owned businesses.

In the absence of policies that substantively promote local proprietorship and cooperative businesses, incremental gains in communal economic activity could be achieved by bringing some businesses back into neighborhoods and out of regional malls. Municipal policy could require that big-box retailers opening stores on the outskirts of cities simultaneously place a branch store downtown. While doing so forgoes the economic multiplier gains of local proprietorship and leaves the store's future subject to the whims and cost-benefit calculations of distant managers, it nevertheless lessens downtown decay and blight. The downtown retail space occupied by large chains would be better maintained by their owners and continue to contribute to municipal coffers. Such a policy is a potentially promising incremental step toward eventually realizing a thicker community economy.

Transportation Systems

Transportation systems could be more communitarian if they promoted walking and supported neighborhood social centers rather than funneling individual drivers onto highways as quickly as possible. The previous chapter already covers many of the relevant urban policies that sustain automobility. Here I focus on the barriers that more specifically block opportunities for improved public transit.

The most significant barrier to improved public transit is financial. Indeed, the early twentieth-century demise of electric trolley lines in the United States was precipitated by the financial consequences of poorly regulated suburban expansion. Municipalities required that trolley companies serve the newly built low-density suburbs without providing much financial assistance. Combined with the Great Depression, these policies saddled trolley companies with huge debt loads, making them easy targets to be bought out and dismantled by a consortium of bus, automobile, and oil companies.[30] Today, "free market" thinking leads some public officials to demand that public transit authorities fund themselves primarily through fares. Economic analysis of such funding arrangements finds that they lead to emaciated public transit systems that mainly serve captive markets—that is, students, the poor, seniors, disabled persons, and others who are unable to afford or operate an automobile.[31]

Although better financial support of public transit is no doubt stymied by the low cultural valuation of mass transit among some policy makers—the aphorism that "a man who, beyond the age of 26, finds himself on a bus can count himself as a failure" is often ascribed to conservative politicians such as Margaret Thatcher—ignorance of transport economics also plays a role. Public transportation is characterized by high fixed costs and very low marginal costs: it costs almost nothing to add another passenger to a bus or train car. Moreover, the addition of more passengers rarely affects the speed of service. Increasing the frequency of service by adding extra buses or cars in response to rising demand is inexpensive relative to the fixed costs of service as a whole. Moreover, doing so produces social benefits through decreased waiting times. Hence, such systems exhibit significant economies of scale. Road networks are the opposite: the addition of cars to a road or highway very soon leads to increased congestion, commuting times, and pollution. Congestion is alleviated—and only temporarily—by laying additional concrete or asphalt at considerable expense. Moreover, service quality usually goes down and social costs ramify as additional users are added to roads or highways. Road networks, therefore, exhibit significant diseconomies of scale.

Thus, it is unsurprising that transportation economists have contended for decades that efficient transportation systems require congestion taxes for roads and the subsidization of public transit. Indeed, Nobel Laureate economist William Vickrey argued that fares should be set as close as possible to the marginal social cost of each additional passenger, which would mean that ticket costs would not be expected to cover infrastructural and other fixed costs.[32] The subsidy of fares and the large fixed costs of transit networks, according to Vickrey, should be recouped by recapturing the increased value of the land surrounding transit stops. The economics of this arrangement, of course, works only to the extent that fare subsidies and tax revenues ensure well-patronized, quality service. Only then does mass transit spur development around transit stops and, hence, increase land values—often referred to as *agglomeration benefits*.

Studies have confirmed that property values tend to increase substantially around light rail and metro stops, though other factors are also influential.[33] Municipalities could seek to capture some portion of this increased value through various mechanisms to fund transit infrastructure, including land value or "betterment" taxes, tax increment financing, and local or business improvement districts.[34] The city of Portland, Oregon, has helped fund its streetcar line through a local improvement district, though exempting owner-occupied buildings from the special levy. Similarly, London is funding a new commuter rail system through a supplemental business tax. When properly targeted, these funding mechanisms help to fund transit progressively. Barring the prior establishment of a community land trust or similar arrangement, of course, areas near transit stops are likely to gentrify. Value-capture schemes, however, can make certain that such gentrification nevertheless results in a well-funded public transportation network that serves the needs of those pushed to live several blocks farther away by rising property values.

In the absence of land-value or hybrid taxation schemes, municipalities can simultaneously support transit while disincentivizing sprawl through the use of transportation impact fees. A case in point is the city of Santa Monica, California, which charges roughly two and a half times the impact fee for single-family homes as for multifamily units.[35] Such policies can help ensure that low-density suburban growth contributes more significantly to municipal transportation budgets.

Another way of supporting transit is by disincentivizing driving. London, for example, charges an eleven and a half pound fee to people driving into downtown during the traditional work week, which has helped fund bike lanes across the city and improvements to the bus system. Another approach is to eliminate the provision of employer-subsidized parking, which often gets classified as a tax-exempt fringe benefit and thereby forces all workers to help pay for car commuters' parking spaces through higher tax rates. Studies have shown that it inhibits transit use and incentivizes solo driving.[36] Even employees offered the option between free parking and free transit passes will tend to drive. Only when firms subsidize only car-sharing or transit—making solo drivers pay full price—do employees drive less.

Finally, the improvement and growth of public transportation systems across North America usually progresses at a snail's pace. Los Angeles's roughly nine-mile purple line extension is not projected to be complete until 2035. Compare this with the metro in Washington, DC, which between 1969 and 1979 grew from nothing to almost thirty-four miles. Despite significant technological advances, public transportation is more expensive and more slowly built than decades ago. Although the above-mentioned alterations in funding could help speed construction, political-cultural barriers also need to be addressed.

Bureaucratic requirements now play a big role in slowing transit growth via requirements for the numerous expensive and time-consuming studies prior to breaking ground; authorities must discern the environmental risks, anticipated ridership, and the broader effects on transportation.[37] Such requirements result in many projects taking up to a decade to move from conception to beginning construction. Some of these studies are no doubt important for wise public decision making and limiting environmental damage. Many requirements, however, introduce unreasonable hurdles to improving public transit. Slowing transit growth with environmental impact studies is self-stultifying when it results in the unabated increase of automobile traffic. Mechanisms should be put in place to ensure that light rail is not built on top of ecologically sensitive wetlands, of course, but the broader environmental benefits of improved transit should afford public transit projects more leeway in such studies than is granted to highway construction.

Requirements for traffic impact studies hamstring transit authorities' efforts by forcing them to ensure that their projects entail almost no risk of slowing the movement of automobiles. One of the most egregious examples is the Washington Department of Transportation's decision to block light rail expansion in Seattle until the transit agency replaces the two lanes on a freeway bridge that its trains will eventually occupy.[38] Like overly rigid environmental study requirements, the Seattle decision misses the larger picture: improved transit invariably means fewer automobiles and less congestion. Moreover, traffic impact requirements fall unfairly on public transit proposals. Highway builders are seldom required to prove that their projects do not undermine public transit or other modes of transportation, namely walking and biking, prior to beginning construction. If they faced the same requirements and restrictions as public transit, it is unlikely that many urban highways would actually get built.

Governmental bureaucracy, however, is not the sole reason for paltry progress in the growth of mass transit. Transit authorities, for their part, tend to propose grand, expensive projects. Given the high up-front costs, it is reasonable to demand ridership projections and other studies; some accounting of public benefit versus cost should be done. In any case, transit planners too rarely propose dedicated bus rapid transit (BRT) lanes as stepping stones to light rail or subways. BRT systems provide similar levels of service at less than half the capital costs of light rail and less than one-tenth that of a subway.[39] Indeed, providing more mass-transit punch per dollar is why such systems are often found in less wealthy Asian and Latin American countries. At the same time, BRT lanes create sociotechnical momentum for later transit improvements: it is politically much easier and less expensive to install overhead wires and steel rail on the right-of-way already established by dedicated bus lanes than to start from scratch. Finally, building BRT lines ahead of planned light rail development provides insight into future ridership possibilities and can encourage earlier densification. Edmonton, Alberta's plan for a city-wide expansion of its light-rail system is proposed to take twenty-five years, a long time for residents to wait. This expansion could be developed on a timescale more along the lines of ten or fifteen years if municipal authorities considered a more incremental pathway via BRT.

Finally, public transportation systems frequently face stiff opposition from conservative suburban dwellers, often under the spurious pretenses of "fiscal responsibility." Cities and states, for such actors, simply cannot "afford" transit, nor should public dollars be used to subsidize the lifestyle preferences of urbanites. Such claims rarely take into account the fact that public dollars already heavily subsidize the lifestyle preferences of automobile drivers, and governmental officials express few budgetary qualms when approving incredibly expensive highway tunnel projects. Consider Seattle's plan to spend billions to rebuild two miles of the Alaskan Way Viaduct in an underground tunnel.[40] Advocates of public transportation, nevertheless, need not focus on such inconsistencies. Rather, they are likely to be more effective by focusing on how improved transit even benefits automobile drivers, and by building alliances with businesses whose bottom lines are hurt by congestion. Consider how Utah, hardly a bastion of liberal politics, has doubled the length of its light rail and commuter rail service (from 63 to 133 miles) in seven years.[41] Advocates pointed out that ridership would eliminate two full lanes of traffic on Interstate 15 and garnered the support of UPS, who advertised to consumers that improved transit helps ensure that their packages arrive on time.

Other strategies employed in Utah reduced the effects of intermunicipal competition and opposition.[42] Transit authorities established early that they would provide the basic infrastructure, while individual municipalities would be responsible for the amenities surrounding commuter rail stops. Moreover, they secured much of the necessary right-of-way from the Union Pacific Railroad back in 2002, eliminating the need to negotiate with individual municipalities. Indeed, not-in-my-backyard opposition to the perceived noise of light rail systems or the efforts by residents in affluent neighborhoods to seek a better deal for themselves—at the expense of their neighbors—can substantially slow the development of transit networks.[43]

Public mass transportation systems no doubt vary in how well they support dimensions of thick community. Apart from encouraging walking, many New York City subway stations do little to center community life. Very often they are nonplaces that busy individuals aim to walk through as quickly as possible. Hence, lessening the barriers to public transit needs to be combined with a positive focus on ensuring that transit stops and stations can double as public gathering spaces. Park-and-rides are certainly not up to the task. Transit access points need to be surrounded by local retail as well as places for leisure, sitting, and people watching to take advantage of already increased foot traffic. Ensuring that important transit stops are designed in this way would require improved planning processes and changes to code. Frequently, bus stops are an afterthought, placed after neighborhoods have already been built rather than explicitly planned for.

Many of the changes recommended in the last chapter—such as improved training of planners and architects as well as form-based codes—could help ensure that expanded public transportation is effective in encouraging community.

Toward a More Communitarian Internet

Similar to the infrastructures discussed above, realizing a more communitarian Internet would require that information networks are at least partly centralized at the scale of neighborhoods. That is, the technology would need to be meaningful for users at social scales between that of individuals and the region or nation. Doing so would enable community-scale governance and result in an Internet that does more to support community-level social interactions and exchanges. Although the current World Wide Web has helped some users find social support and has aided the coordination of economic charity and social movement organizing, it could still be more communitarian.

One particularly promising alternative is the community *mesh network*.[44] Such networks are composed of a multitude of wireless hotspots located within each other's line of sight. The geometry of mesh networks strongly contrasts with the arrangement of backbones and central servers that make up contemporary Internet infrastructure: they are *ad-hoc* networks that function as long as a sufficient number of nodes are up and running. Several experimental community mesh networks are currently running in Europe and countries like South Africa. The largest are Guifi.net in Spain (more than 20,000 nodes) and the Athens Metropolitan Network (more than 2,500 nodes). Maximizing the communitarian benefits from mesh networks would require ensuring that they are not simply anonymous gateways to the Internet; they would need to offer users something of their own. Mesh networks could support a range of community resources, including access to municipal services, neighborhood message boards, connections to local businesses, and communication tools like video chat.

The wider implementation of community-scale mesh networks face several barriers. First, there are technical barriers needing more research and experimentation.[45] Coordinating the movement of information through ad-hoc networks, since they lack central servers, becomes more difficult as the number of nodes and traffic increases. Most current implementations of mesh networks, as a result, tend to be not more than a couple hundred nodes. This barrier could be partly lessened by taking advantage of the *dark fiber* networks that many cities laid between municipal buildings years ago but never used.[46] Municipal or public school networks could form the backbone for larger mesh networks or facilitate the communication between neighborhood level initiatives, limiting the effect to which high levels of traffic harms service quality.

Legal challenges could also limit the advance of mesh networks. The Communications Assistance for Law Enforcement Act (CALEA) in the United States requires that "telecommunications carriers" deploy their hardware and service such that communications can be surveilled by law enforcement. It is not clear if police surveillance is possible within current mesh networks, except when users access the Internet outside the network. While community mesh networks have yet to be targeted by legal action under CALEA, they might be in the future if legislation is not passed to exempt them. Established Internet service providers (ISPs), for their part, are likely to lobby to quash any potential competition from mesh networks. For instance, they might prohibit customers from letting mesh-network users access the Internet through their connection, perhaps employing squads of workers to detect and report any user acting as a gateway to the Web for others. Community mesh networks might best prepare for these challenges by partnering early on with municipalities or anchor institutions large enough to weather legal challenges and negotiate with ISPs.

Another potential barrier stems from the prevailing culture of Internet activists. Advocates of online communication technologies tend to view information as ideally free, both monetarily and from overt forms of social control. Indeed, most current mesh systems neither charge for nor track users' usage, which presents inevitable limits to growth. The supply of bandwidth provided by altruistic users can eventually be surpassed by the costs of hardware and expertise needed for maintenance and expansion. Likewise, barring other means of compensating the users who offer up their Internet connections, such as esteem, they may become frustrated with the resulting loss of bandwidth and shut off their connection with the mesh network. The lack of structures for effectively coordinating and governing infrastructure and usage puts such networks at risk of being undermined by "tragedy of the commons" situations.[47] In any case, despite democratic appearances of voluntaristic mesh networks, they are perhaps better thought of as benevolent dictatorships: they function not through collective governance but only as long as a handful of people believe it is worthwhile to donate their ISP connections.

Mesh network governance structures could be more democratic than anarchic or libertarian by enabling communities of users to self-regulate usage as well as decide how to charge for access to the broader Internet. Such charges would no doubt need to be lower than an ISP subscription if mesh networks are to attract users. In any case, being able to track download rates of users, among other monitoring capacities, would be necessary to effectively govern the network. Indeed, exactly just such a monitoring system was implemented in a rural community mesh network in Wray, England, to prevent an informational commons tragedy.[48] Internet users are likely to shy away from such arrangements, however, as long as they view freedom as simply the absence of overt oversight and regulation rather than something that can be achieved through the practice of democracy.

Any future for mesh networks will be shaped by the outcome of current political disputes over net neutrality.[49] Opponents of regulatory changes that enable ISPs to "throttle" access speed to or slap higher access charges on bandwidth-hungry uses, like movie streaming, fear that such changes will "break the Internet." Internet nonneutrality is believed to lead to the degradation of connection quality, to ISPs providing the highest access speeds only to the websites and services that they have a financial stake in, and to ISP "censorship"—the throttling of access to views that ISP owners disagree with. While such consequences are possible, they are likely not the reason that large Silicon Valley firms have taken such an interest in net neutrality. Indeed, the controversy can also been seen as a political skirmish between ISPs and large corporate content service providers, such as Netflix, Google, and Facebook, who rely on the current Internet regulatory and pricing regime for their business models. Rather than a battle for freedom writ large, the dispute is really over which set of corporations will be in the most privileged position to profit from the Internet.

Given the current Internet's colonization by large corporations, might it be better to let ISPs "break" it? On the one hand, a "two-tiered" Internet might be eminently desirable, for instance, if top-tier advantages were not awarded to those who simply were more able to pay but to community networks. On the other hand, it may be desirable for citizens to cut their losses and pursue alternatives. Former Internet cheerleader Douglas Rushkoff, for one, has argued that the Internet was never truly bottom-up or "free" and has proposed that we "abandon" it.[50] Hence, those favoring a more communitarian Internet might do well in doing nothing. Neglecting to shore up the current instantiation of the Web could help drive interest in pursuing and building alternatives.

More Communitarian Organizations

Similar barriers are faced by communitarian organizational structures, which coordinate the provision of services such as physical recreation, spirituality, or schooling. Some of these stem from the advantages of scale and centralization. Budget crunches provide significant pressure to centralize. Four-rink hockey complexes and equally gargantuan wellness centers afford savings in management costs and other operational efficiencies. Nonetheless, similar to the case of big-box retailing, this is done by externalizing costs onto users. Greater recreation center economies of scale come from forcing patrons to drive. Rarely is it recognized that increased centralization undermines the potential of such places to be community-centering institutions. Decentralizing recreation centers might be accomplished by revising accounting practices. Many large leisure centers likely measure and justify their efficiency in narrow economic terms: the number of dollars spent per visit. Attaining narrow economic efficiency in such ways, however, inevitably drives away users who dislike traffic or live closer to private alternatives. A more systematic accounting system for recreation centers would include expenses externalized to users and measure the extent to which the institution centers community life (e.g., percentage of local neighborhood served).

Similar economies of scale encourage municipalities to build schools that hold thousands of students. Such scales, however, make them function and feel more like large, impersonal bureaucracies than communities. Some parents have likely experienced PTA meetings where "participation" means being herded into gyms or large auditoriums to hear talks from administrators. In the short run, little can be done to alter already massive school buildings and complexes. Yet large schools could be virtually broken up into smaller institutions. The barriers to doing so are mostly people's imagination and political willpower. Many school buildings in American inner cities are already subdivided into small learning communities, and some studies suggest that the resulting personable and communal learning environment can improve academic achievement among poor minority students.[51] Bringing the scale of public schools down to a personable level would also enhance the potential for political community among teachers, administrators, students, and parents.

Cooperative or democratic schooling would be the thickest option, in terms of political and psychological community. The bureaucratic or quasi-authoritarian approach to schooling, however, has a substantial degree of momentum or obduracy. The industries profiting from curriculum development, educational consulting, and standardized testing have a vested interest in the current system. People's thinking is similarly limited by their adherence to dominant frames wherein learning is something that happens only when state governments carefully dictate its pace and structure. In the United States especially, the language of "accountability" justifies standardization via intense testing regimes and the subsequent elimination of all noncompatible forms of pedagogy.[52] In turn, many schools can seem more like assembly lines of lecture and Scantron sheet regurgitation rather than places for deep learning. Moreover, parents with anxieties regarding their children's futures in an age of increasing job insecurity are liable to believe that squeezing more and more lecture and testing into the school year will invariably produce smarter and more successful adults.

These dominant ways of thinking stand in the way of more democratic schooling, especially the belief that only through heavily bureaucratized institutions and standardized curricula are students adequately prepared to begin their journey as ladder-climbing rugged individuals. Studies of democratic school graduates challenge such beliefs, finding students to be no less successful than their fellow middle-class but bureaucratically schooled counterparts. Indeed, many go on to earn PhDs, become professional musicians, or start their own businesses.[53] Nevertheless, what "success" means depends on whether one embraces the moral imaginary of networked individualism or more communitarian visions. Some people see the matter very narrowly: success means being credentialed by a top college or university and securing an elite position within the global job market. Under these terms, prosperity accrues primarily to individuals—communities are mere stepping stones. Must success necessarily mean leaving one's community rather than filling a need within it or working to make it better?

Despite these cultural barriers, more democratic or cooperative schooling could be incrementally steered toward simply by enabling more experimentation with the model. Indeed, technological societies need not convert all their educational institutions into democratic schools—like the Sudbury School—to realize some of their benefits. Just as Google encourages workers to spend 20 percent of their workweek pursuing projects of their own interest, schools could be encouraged or required to permit a similar amount of semistructured and collaborative exploration. Alternatively, such time might be used to discuss class rules and work through interpersonal conflict. Children might use this time to organize activities with younger cohorts, help run a school garden or animal farm, or engage in self-structured play. Integrating democratic approaches into teacher training programs, moreover, would likely be necessary to ensure quality experimentation, and state-level curriculum requirements would need to be revised.

At the same time, the rarity of Sudbury-model and other "free schools" is likely to lead many people to see it as infeasible or impractical, perhaps working for the children of affluent counterculture parents but not for others. Though regulatory capture by the testing and textbook industry is likely to remain a barrier, the U.S. Department of Education could lead a systematic testing of the democratic school model by outlaying funds for several districts to experiment with the model in one of their local schools. Such a study would necessarily take decades to provide comprehensive results that could allay parental anxieties about the feasibility of such schooling models but, in the meantime, would at least ensure that a more significant portion of Americans have experienced community in an educational setting.

The barriers to realizing thick community in recreation are less daunting. Indeed, given that bureaucracy typically makes recreational activities more expensive, self-organization would result in more affordable sporting opportunities. Tenacious cultural and cognitive barriers, however, especially among parents, are likely to stand in the way. Many adults today impose professional standards and criteria onto youth sports, forgetting that paid athletes focus solely on performance in part because millions of dollars of ticket sales and advertising revenue, and hence their paychecks, depend on it. Overemphasis on scores, batting averages, and other measures leads to a displacement of opportunities for play, which is further exacerbated by parents with high aspirations for their children's ostensible sporting futures (e.g., college scholarships). It is easy to lose sight of the value of experimentation and conflict negotiation to children's development when the point of youth sports is presumed to be preparing children for the professional leagues. Even worse, some parents model violent, antisocial behavior at youth sporting events that is the antithesis of political community, most notably in extreme cases such as when an angry parent has beaten another to death.[54] Performance-oriented thinking about youth recreation further synergizes with cultures of fear regarding child predation to decimate opportunities for self-organized outdoor play.[55] Finally, the flight from political community in sporting activities is exacerbated by what Theda Skocpol described as the shift from an ethic of membership to one of management—that is, the belief that public goods should be produced *for* citizens *by* professionals and specialized experts rather than *through* and *alongside* citizens.[56]

Mitigating these cultural barriers to developing more communitarian organizations is no simple matter. It is far more straightforward to implement a bureaucracy than to lay the groundwork for groups of citizens to better help themselves. Little League clubs and traditional schools seem easier to institute than providing the conditions for sandlot baseball and democratically run schools. How can governments and citizens better organize for self-organization? Doing so for youth sports would entail treating them more like community swimming pools than miniature professional teams. While youngsters do take swimming classes and join competitive swim teams, one rarely finds people under the age of eighteen swimming laps at the average municipal pool. Rather, children are engaged in self-structured play under the minimal supervision of lifeguards. Municipalities could broaden participation in sports, while also allaying parental fears about "stranger danger," by paying teenagers to hang out in local sporting fields encouraging play. Sports "lifeguards" would be tasked with intervening to prevent major injuries and making sure bats and balls do not disappear. They would participate as a player, but rarely as a coach or an umpire.

Developing more communally oriented churches and other spiritual organizations also faces entrenched cultural barriers. Some citizens have embraced an extreme form of religious individualism,[57] while others have become very skeptical of religious metaphysics and metaphor—believing that a scientific understanding of the universe is incompatible with spirituality. Such cultural patterns limit the possibilities for more communitarian religious life. The emphasis on maximizing one's personal return on spiritual investment via self-actualization within religious individualism discourages collective involvement; a strict utilitarian logic makes less conditional commitments to one's fellow worshippers seem irrational. As a result, the religious individualist's search for answers to "big questions" can come to look more like consumerism than communality. On the other hand, religious skeptics, who tend to view spiritual texts as trying to provide epistemological explanations of the origins of the universe rather than as metaphors, forgo an embodied connection with a spiritual collectivity for a more diffuse sense of identification with other "free thinkers." An unwavering insistence on proof, as opposed to a willingness to entertain faith, doubt, and uncertainty, leads many skeptics to sacrifice religious communality for the pursuit of a more "rational" understanding of being. Skeptics and religious individualists, as a result, let a narrow focus on *belief* get in the way of *belonging*.

Given the growing dominance of religious skepticism and individualism in many technological societies, the solution for religious institutions probably lies in learning to leverage spirituality into new (and more civil and inclusive) ways of talking about and supporting togetherness. Overemphasizing sectarian differences, being stubbornly inward looking, and failing to productively engage with doubt can make even the most progressive church seem conservative to people who are a generation removed from regular religious service attendance. Theologians and religious practitioners, of course, have been thinking deeply about these challenges for some time;[58] nevertheless, it remains uncertain if more communally engaged *emerging church* models will be able to gain momentum over the *gated community* model of many strains of evangelism.

Policy changes could encourage churches and other spiritual organizations to be incrementally more thickly communitarian. One approach is relatively straightforward: reform the laws relating to tax-exempt status. Currently, tax exemptions are uniformly extended to nonprofit organizations, in the United States and around the world, regardless of their public effects. A church whose mission includes running a soup kitchen may receive the same tax exemption as a real estate lobby group wanting to "clean out" a homeless population in order to improve sales of nearby luxury condominiums, despite the latter's more tenuous relationship with most notions of the public good. The Boys and Girls Clubs of America receives much the same class of tax exemption as a private university that charges fifty thousand dollars a year in tuition and builds posh dormitories for upper-class eighteen-year-olds.

Tax law for corporations and nonprofits could be reformed to contain several different tax categories. A public subsidy for a set percentage of an organization's operating expenses would be provided only to organizations that offer free or below-market cost services to lower-class citizens within their neighborhood's borders. Such services, moreover, would need to exceed some large minimum threshold of the organization's budget. Institutions like soup kitchens and food pantries would be most likely to fulfill these conditions. Though not providing communal benefits, organizations providing medical assistance to poor nations in the global south might also be included in this category via another set of criteria. Nonprofit organizations for whom community service composes a much smaller part of their mission, in contrast, might pay somewhere between no taxes at all and only at a slightly reduced rate, depending on whether its main clientele and beneficiaries are primarily upper class. Also contained within this tax bracket would be more insular churches, whose budgets simply ensure that a building is available for Sunday worship and that a pastor remains on call for congregational needs—that is, those that neglect to build community ties with nonmembers or neighbors. A food co-op might pay some tax, if it catered explicitly to middle-class foodies, or no tax at all, if it used groceries nearing their spoilage date to make hot meals for the homeless and others in need. Another variable in the revision of tax-exempt status might be the presence or absence of political community. A nonprofit run autocratically by a wealthy philanthropist could be required to pay taxes at one-quarter to one-half the rate of a for-profit firm but nothing at all if managed collectively by employees and other stakeholders.

Similar tax-incentive systems could exist for for-profit companies. *Benefit corporations*, firms that are not legally obliged to single-mindedly maximize shareholder value, could have their tax burdens reduced to the extent that they sacrifice profits for community supportive ends. For example, Hobby Lobby might receive a half percent tax break for its policy of not opening on Sundays, which is meant to permit all workers the ability to spend time with friends and family. Larger reductions could come with worker ownership, paying a living wage, having local community leaders sit on the company's governance board, or locating work spaces closer to workers' residential neighborhoods.

I have outlined a multitude of barriers that would need addressing to reconstruct technological societies to contain more thickly communitarian infrastructures and organizational technologies. The lack of specific policies for dealing with community-scale energy infrastructures amplifies uncertainty, adds bureaucratic delays, and increases the costs of such systems. The failure to limit smokestack chasing advantages large corporations and harms local proprietors. Potential synergies between consumer cooperative structures, unions, and anchor institutions remain underrealized. Poorly designed tax regimes and fare-setting systems stymie the development of public transportation. Internet activists devote too many dollars and far too much time fighting to shore up a weakly democratic, thinly communitarian, and corporate-dominated World Wide Web when they could commit their energies to building alternatives. The insistence that information be costless and free of social control, furthermore, hampers the effective governance of bottom-up Internets. Recreation and community centers would more often be scaled to support local webs of social ties if public accounting did not permit the passing off of transportation costs to users. The barriers to youth sports and schools that are less bureaucratic and thicker in political community, on the other hand, have more to do with thinking than finances. The fetishization of performance and accountability as well as an inability to think outside an assembly-line model of education blocks the development of more opportunities for less structured play and more democratic schools.

There is nothing natural about the status quo of highly centralized and bureaucratic infrastructures and organizations. This status quo is the product of a constellation of sociopolitical and psychocultural factors that have accumulated over time and become obdurate. Reconstructing technological societies so that citizens see themselves not as individualized operators of vast sociotechnical networks but as embedded within more stable and localized webs of communal relationships is a matter of incrementally altering or eliminating these factors. If more citizens were to gain experience in democratic schools, cooperative businesses, community-governed mesh networks, and neighborhood-scale energy improvement districts, at least some of them surely would come to recognize that highly centralized networks are no more inevitably "the future" than suburban sprawl and anonymous high rises.

Nevertheless, simply building infrastructures that call for cooperation and local-scale social interaction is probably insufficient. Many citizens are habituated to networked or privatized living long before having the chance to participate in a community energy system or food cooperative. Similar to how new urbanist planning and design is unlikely to automatically turn those already disposed to being wary of getting "too involved" with neighbors into convivial communitarians, the presence of more communal infrastructures is by itself insufficient. In the next chapter, I explore the barriers to altering the techniques and artifacts that, much like Little League and bureaucratic schooling, leave citizens poorly prepared to pursue thick community.

9 Achieving Communitarian Techniques and Artifacts

Much of the research on the shaping power of technologies and the barriers to technological change tends to focus disproportionately on structures like urban form, missing the more subtle influence of other technical features of contemporary societies.[1] Everyday techniques and artifacts could better help develop citizens that are more disposed and prepared to act communally. Changes to these technologies are at least corequisites to achieving thicker communities. What are the barriers to alternative gadgets, tools, and child-rearing techniques, and how could they be lessened?

The barriers I discuss are somewhat different from those for urban and large sociotechnical systems. Infrastructures tend to be large, expensive, and ultimately reliant on some form of public or collective coordination. Although televisions or child-rearing techniques inevitably depend on communication networks or systems of pediatric information and authority to function, they are implemented to a much more significant degree by their users. That is, users "choose" to purchase and locate a television in their home or to acquire a concealed-carry handgun license (CHL). In what ways would the decision landscape faced by users need to be altered to enable them to make more communitarian choices? What policies, infrastructural and organizational changes, and realignment of incentives might be necessary?

Toward a More Communitarian Domestic Environment

In chapter 6, I argued that domestic environments better prepare citizens for the practice of community when they encourage social gathering and interaction rather than dispersion and isolation. Home heating is one of the most significant shapers of domestic environments, at least in nations with a cold winter season. Homes would better encourage domestic gathering if they were heated by centralized point sources rather than by distributed "forced air" or boiler systems. The point source need not be a woodstove, which produces soot and may aggravate respiratory ailments, but could be fueled by almost any type of energy: natural gas, wood pellets, or electricity. Air- or ground-sourced heat pumps could easily be designed to terminate in one or two large radiators or central fans. Moreover, even passive solar home design can provide a comfort incentive toward congregating by locating the main living space in rooms with the most wintertime sun exposure. Put bluntly, the barriers to more communitarian home heating are not technical; suitable alternatives already exist.

Might the barriers be financial? Installing ground-based heat pumps or converting a home to reflect passive-house standards undeniably entail significant up-front capital costs that might take a decade or more to recoup. Publicly subsidized low-interest loan programs for such technologies would need to be expanded if such approaches are to become feasible for a greater portion of homeowners. The financial barriers to replacing old furnaces with centralized gas/pellet stoves or air-based heat pumps, in contrast, are not so significant. Indeed, the upfront costs of buying and installing such alternatives are often no more than that of a new centralized furnace. Moreover, simple fans for pushing warm air between floors and rooms, to prevent bathrooms from becoming frigid, are less than a hundred dollars and can be installed by even minimally competent do-it-yourselfers.

That more homeowners have not sealed their distributed HVAC systems shut and installed some nondistributed heating unit appears to be mostly the result of ignorance as well as entrenched practices and patterns of thought. Forced air systems are frequently perceived to be more "modern" than centralized heaters or stoves.[2] Many home dwellers have become habituated to sleeping in seventy-degree bedrooms in the winter, and hence might find wearing pajamas or sleeping with heavy blankets to be an adjustment. People with compatibly arranged homes have likely never even considered replacing their forced air system with a centralized stove. Offering tax credits for these heating alternatives, as is already done for high-efficiency furnaces, could help lessen the effect of the relevant cultural and cognitive barriers. Given the role of centralized heating technologies in promoting domestic togetherness, champions for these tax credits might be easily found among those elected officials whose image relies on the promotion of "family values."

Ensuring more communitarian thermal environments in new construction would demand changes at multiple levels of the building industry. As sociologist Elizabeth Shove has argued, contemporary architectural science has come to idealize a standardized conception of comfort that ignores cultural and geographic differences.[3] Architectural science tends to portray the uniformity and stability of temperature offered by complex climate-control systems as unequivocally ideal; natural patterns and other forms of thermal contrast become problems rather than features. Architects, of course, are not solely to blame for this state of affairs: they are constrained by building codes, dominant cultural expectations, and the designs, practices, and standards of heating and air-conditioning engineers. Insofar as such understandings of thermal comfort, as well as their ancillary standards and codes, continue to predominate, new homes are unlikely to include features such as centralized heating elements.

Real estate agents and developers, moreover, can erect their own barriers. They tend to err on the side of conservatism, because homes that deviate too significantly from the status quo represent a financial risk. Certainly the above-mentioned tax breaks for heating systems that better promote domestic community would entice home builders to include them more often in their designs. Nonetheless, they are likely to be more responsive to changing consumer preferences, which could be nudged by new furnace labels. These would be similar to the EnergyGuide labels that provide an estimate of an appliance's yearly gas or electricity usage, instead describing its likely social effects. One might read "Warning: Forced air heating systems are designed to distribute heat fairly uniformly within homes and buildings. Such systems have been found to encourage social privatism and lower amounts of interaction among household members."

Ensuring that more domestic environments promote gathering over dispersion may require discouraging a range of other technologies, including personal electronics and individual microwave dinners. Although thick community is probably best supported when television is not central to domestic leisure, discouraging households from having more than one TV would be an incremental improvement. The isolation of household members within their own media cocoons benefits advertisers at the cost of preparing citizens for political community. Numerous countries in Europe, Africa, and Asia already charge residents television licensing fees. If these fees were scaled with the number of televisions, such countries could discourage televisions in residents' bedrooms at the same time that they raised funds for public television and other collective goods. Instituting such fees in North America, where there is little history of television license fees and broader cultural opposition to taxation, would likely be infeasible. Nevertheless, regulators might more deftly sidestep popular opposition to influence consumer behavior by attaching graduated licensing fees to the set-top boxes needed for cable or satellite access.

Policies designed for television and set-top cable or satellite boxes, however, may become obsolete and ineffectual as more and more viewers shift to streaming video services like Netflix, Hulu, and Amazon Prime. For those with basic high-speed Internet service, streaming two movies simultaneously is often infeasible, which forces household members to negotiate over what to view and when to view it. As Internet infrastructure improves in countries such as the United States, however, the need to negotiate over what to watch on Netflix might someday be technologically obsolete. If Internet service providers were allowed to differentially charge for content access based on the type of content, then policy makers could use disincentives to steer viewer behavior. The cost of residential Internet access would float based on type of use. Downloading multiple video streams on a single connection, which signals domestic dispersal, could be made to cost more. Such policies, however, are likely to be infeasible insofar as citizens consider a "neutral" Internet to be an almost natural right.

Domestic environments are more communitarian, moreover, when meals are collective rather than solitary affairs, regardless of whether a household is made up of family members or roommates. Individualized microwaveable dinners and similar foodstuffs help support a more networked than communal household. Using policy to target microwaves would probably be too blunt of a response, given that they are frequently put to use in preparing collective meals. Instead, policy makers could tax individual-sized microwaveable meal and similar processed food items, using the funds to subsidize basic staples and fresh produce to incentivize cooking. Doing so would be a significant reversal of current American farm policy. Indeed, some 20 billion dollars in federal spending since 1995 has gone to subsidize the production of corn-based sweeteners and thickeners as well as the soy oils used in processed foods like TV dinners. In direct contrast to the USDA food pyramid, federal dollars go primarily to dairy and meat producers—partly via corn and soy subsidies while growers of other grains, nuts, and legumes; fresh fruits; and vegetables get a measly 13, 2, and 0.4 percent of the available subsidy dollars, respectively.[4] EBT, or food stamp, dollars could count double for fresh produce and other staples but not for frozen dinners. Indeed, in many American states, food stamp dollars spent at farmers' markets are matched one-to-one, permitting those getting welfare to pay half price.[5] Expanding such policies would promote purchasing staples over processed foods more broadly.

On the other hand, reversing this subsidy pattern may do little for people who are short on time and money. People working multiple jobs or long hours to stay financially afloat can hardly be blamed for lacking the energy to cook a big meal at the end of the day. Policies directed toward shortening the work week, raising minimum wages, and working to keep housing costs low would have significant influence on the feasibility of domestic meal times. Ironically, conservative politicians tend to fervently oppose these policies, even though they would do quite a lot to promote family values. Some of those politicians, in any case, might be turned into allies if leftist advocates more often described such policies as enabling family life, rather than as only promoting social justice.

Citizens are also encouraged to buy frozen and other prepared foods by the diffuse distribution of large supermarkets, enforced by suburban building patterns. Buying fresh, and hence quickly spoiling, ingredients for preparing meals requires going to the grocery store at least once a week, if not more often. Insofar as that task is a chore made more onerous by several miles of traffic, competing for a parking space, and navigating a massive and crowded supermarket, consumers are likely to minimize how often they do it. Denser mixed-use development patterns would enable residents to more easily choose thick communitarian mealtime practices.

Other factors make frozen foods cheaper than they otherwise would be. The near complete dominance of frozen foods in school lunch programs subsidizes the industry's economies of scale. Lobby groups such as the American Frozen Food Institute and the National Potato Council have fought tooth and nail against proposals to impose limits on salt levels and starchy vegetables in children's lunches or requirements for an increase in fresh fruits and vegetables. Congress even went so far in 2011 as to debate whether the tomato paste in a slice of reheated frozen pizza could be classified as a serving of vegetables.[6] Individual school districts are, moreover, highly restricted in their power to shape their food purchasing. The National School Lunch Program in the United States forces participating schools to shape their meal offerings around purchases of surplus agricultural commodities, typically dairy, certain meats, corn, and soy and in the form of frozen meals.[7] Orders of ingredients, if schools want federal reimbursement, cannot simply be placed with local farmers but must use the USDA as an intermediary between them and large agribusiness conglomerates.

Many schools have experimented with shifting lunches away from frozen prepprepared foods and toward cooking meals on-site. Indeed, some French school chefs already manage to create healthy options for around three dollars a meal by making sure nothing goes to waste.[8] Nevertheless, it is difficult for many school boards to justify even slight increases to the single digit percentage of their budgets going to food service, much less the cost of repurchasing cooking equipment that no longer exists in their "kitchens," unless current subsidies and school lunch regulations are changed. Many school districts, however, do have a Farm-to-School program in place, sourcing some of their food from nearby farms.[9] Such programs could be expanded and subsidized by the National School Lunch Program. Doing so would help shift the food industry and economy-of-scale advantages away from frozen food producers and distributers as much as it would stimulate more communal agricultural economies.

Barriers to Rearing More Communitarian Children

Child-rearing techniques could more often impart communal skills, lessening the degree to which citizens' development is shaped by the myth of rugged individualism. The insistence that children become "self-reliant" sleepers as infants via crying it out in a separate nursery helps maintain the Western cultural idealization of privacy and independence. The emphasis on making infants and toddlers more independent, ironically, parallels a loss of autonomy among older children. Many parents prohibit their children from walking to school alone or even from leaving the front yard, in contrast to previous generations of kids who might have only needed to show up by sundown. Children's practice of autonomy is less often ensured by a thick community of nosy neighbors and attentive strangers, being more frequently exercised within communications, consumer-retail, and automobile networks. Large deficits in public trust and perceptions of "stranger danger" mean that many children's play takes place mainly in the context of play dates or other highly structured, adult supervised activities, limiting their opportunities for unstructured exploration. Although structured activities offer parents an occasion to socialize and interact, they nevertheless tend to deprive children of the chance to collectively self-organize. In any case, play dates teach youth that one socializes through infrequent networked affairs, not through thick community.

The continuing cultural dominance of the nursery and the cry-it-out method over cosleeping arrangements stems in part from the momentum of outdated pediatric "wisdom." Early pediatric science justified separate sleeping arrangements with the presumption that ideal human sleep patterns entail being totally undisturbed for as long as possible, an ideal upset by the presence of other human bodies.[10] Such one-dimensional thinking about human sleep needs has been undermined by the recognition of the role of oxytocin and other hormones related to physical touch as well parental proximity in regulating infant breathing and heart rate. Despite advancements in thinking about infant development, however, archaic traditions can persist. This is likely old news to new mothers who have had to deal with older relatives that stubbornly hold onto antiquated ideas about infant care.

New purportedly scientific ideas about the "best practices" of infant care have emerged to shore up traditional cultural biases. The American Academy of Pediatrics (AAP) has adamantly discouraged any form of bedsharing among infants and parents due to a perceived connection of the practice to sudden infant death syndrome (SIDS)—though they do encourage room sharing.[11] As some sleep scholars have pointed out, existing studies are not as unequivocal as the academy presents, which suggests that AAP recommendations are as much rooted in Western cultural beliefs as in scientific data.[12] Studies that might seem to demonstrate bedsharing to be hazardous frequently fail to disentangle the multitude of risk factors for SIDS, including smoking, alcohol, and drug use, parental fatigue, bedding material, socioeconomic class, and whether the baby is placed prone or supine. Moreover, they usually fail to explain why SIDS is almost unheard of in Japan and other Asian nations where cosleeping and bedsharing is the norm. The AAP's blanket disapproval of cosleeping mainly serves to keep some parents in the dark regarding how to bedshare or cosleep safely and effectively.

Even if supplied with information about cosleeping less undergirded by twentieth-century Western cultural biases, many contemporary parents would have difficulty practicing it. Smaller barriers include altering sleeping environments to be safer for infants by using firmer bedding material, placing mattresses directly on the floor, or finding an appropriate cosleeping bassinet. Better guidance on the range of sleeping practices and appropriate technologies could be offered to consumers in stores that sell bedding. Parents who cosleep with their children, moreover, would have to think more creatively about when and where to meet their needs for private sexual intimacy. More broadly, alternative sleep schedules and practices are made more difficult by already established institutional rhythms and practices. Children from cosleeping households may struggle in Western daycare settings, where independent napping is more of an expectation. For nap time not to be a source of conflict and frustration for all parties involved, daycare workers would need training concerning cosleeping children, perhaps learning from Asian countries. Indeed, workers at Japanese daycares gently stroke and rhythmically tap the backs of babies and young children, even going so far as to lie down next to the child to provide body heat and physical contact to help them sleep without the presence of their parents. Although this practice might seem at first to be excessively labor intensive, and hence expensive, Japanese workers have learned to lessen their workload by getting the assistance of children too old for nap time.[13]

Moreover, stay-at-home parents may find cosleeping unappealing because they already must care for their children during most, if not all, of their waking hours. One can hardly blame them for wanting some modicum of time and space to themselves. Hence, targeting the structure to every-day life that renders stay-at-home parenting a full-time and largely solitary activity is likely a major precondition to more cosleeping. Policies to encourage walkable neighborhoods that promote webs of local ties and have higher densities of third places and other amenities could enable residents to more often parent side-by-side, reducing the overall workload of child rearing—or at least providing the pleasant distraction of adult company.

Inflexible work schedules also make alternative sleep arrangements challenging. It is unsurprising that parents will find the need to tend to children at night anxiety inducing if it is inevitably disruptive to their work lives. Although many developed nations, apart from the United States, mandate and publicly support parental leave, most people face a dichotomous choice regarding employment: work full time or not at all. The forty-hour (or thirty-five-hour) work week remains the near ubiquitous norm. Adjusting leave regulations to allow parents more flexibility in exactly how they partition their time seems like an obvious fix. Why not use public support to allow parents the ability to transition from their leave to a five- or six-hour workday for several years? The barrier to doing so is mainly cultural, though small businesses might find the staffing challenges particularly onerous in the absence of aid or additional incentives. It nevertheless remains entrenched in the thinking of people in affluent nations that work somehow cannot be accomplished without appearing busy for eight hours or more each day.

At the same time, research suggests that the perceived stressfulness of cosleeping has more to do with cultural expectations regarding sleep than with the practice itself.[14] It is the degree to which parents imbue with anxiety any sign that their child has not yet learned to "self-soothe" that is correlated with sleep problems and stress. That is, high expectations for independent infant sleep is a bigger factor in sleep-disturbance-related stress than interrupted sleep per se. Culturally rooted anxieties about sleep render the physical jostling or noise of another person more bothersome than they otherwise would be. Indeed, sleep disturbances and related parental stresses are rarer among Japanese families with young children, despite high rates of cosleeping.[15] The Western tendency to quickly pathologize any sleep pattern that deviates too strongly from the norm of eight uninterrupted hours introduces a counterproductive degree of fear, stress, and anxiety into infant sleep training.

The high valuation of control in Western cultures erects additional barriers to cosleeping. Constantly trying to actively "fix" sleep disturbances can quickly become exhausting and demoralizing for parents. Spending a long night tending to a fussy child, as a result, can feel like something the CIA would do to suspected terrorists. My point here is not to blame parents if they cannot make cosleeping "work." That would be both moralistic and insensitive. Like anyone else, they are *cognitive victims of their culture.* Their cultural milieu has instilled in them psychological dispositions and thought processes regarding sleep practices that are incredibly difficult to overcome. Although more flexible work hours and more visible guidance concerning compatible techniques and technologies would certainly help parents cosleep more often, the necessary psychocultural changes are likely to be ploddingly slow.

Accommodating more communitarian childhoods after infancy would require making more room for unstructured play and communally accomplished autonomy. The cultural reproduction of "stranger danger" beliefs, however, stands in the way. Parental worries about dangerous streets and child abductions by strangers are often cited as reasons for keeping children locked up at home, despite research demonstrating that children are more likely taken or abused by people they already know. Moreover, the rate of injury from domestic hazards are often higher than from those found in the neighborhood. Many believe, contrary to available statistics, that children are safer traveling at seventy miles an hour in the backseat of their parents' minivan than playing in a public park.[16] Most tragic is the trebling of eleven-year-old British children suffering serious injury or death due to accidents with automobiles on the way to or from school, eleven being exactly the age the vast majority of them receive their first cell phone.[17] Parents seeking a sense of security by giving their child a mobile phone can put him or her in greater danger rather than less.

Further stymying more thickly communitarian parenting arrangements is the cultural tendency to place responsibility for children's well-being almost entirely on parents. The question of "Where were the parents?" is frequently the response to an injured, abducted, or dead child rather than a collective call for traffic-calmed streets and better supervised public spaces.[18] Child safety has become an individual-parental responsibility rather than a collective good. Recall the cases in which parents have been arrested on neglect charges for letting their child play in a public park by themselves. Calls to law enforcement over free-range children obliges all parents to keep their kids on a short leash.

Growing perceptions of stranger danger, at any rate, have been partly manufactured by doom-and-gloom news reporting and media frenzies over disappeared and hurt children. As is the case with perceptions of crime more generally, the chasing of ratings through fear-oriented news stories exaggerates risks by amplifying viewers' perceptions of the frequency of everyday hazards.[19] As a result, parents come to believe that child abductions and fatal playground injuries are more common than they actually are. Direct legislation of reporting practices, at least in the United States, is often difficult, given the traditional hesitancy to regulate the media. In the same way that the "fairness doctrine" at one time forced U.S. radio and television stations to present two sides to every issue, however, the effect of "media panics" might be partly mitigated by requiring news outlets to also present statistics on the extremely low probability of any crime or injury affecting children. Such a rule, nevertheless, would do little to negate the effect of fictional television programs such as *Law and Order*, which also shape viewers' perceptions of the reality of crime.

Because it may be infeasible in the short term to alter parental perceptions of risk, a more productive strategy might be to develop and publicly fund "adventure playgrounds" modeled after the Land in Wales.[20] Municipalities could be enticed to replace much of the sterile, some might say unimaginative, playground equipment that was installed over the past decades in response to litigation fears with semisupervised play spaces full of hand-powered building tools, junk, and other "hazardous" things that delight children. Play workers would be trained to instigate play rather than regulate it, intervening if injury looks to be imminent or to provide first aid. Contrary to what one might expect, the accident rates for adventure playgrounds are often substantially less than those of traditional play spaces.[21] Regardless, the mere presence of supervision is likely to get some citizens to relent and allow their children to meet and play with friends without parental surveillance.

Several policy changes would help eliminate some of the barriers to adventure playgrounds. To begin, despite their lower accident rates, required risk assessment paperwork can be burdensome; making changes to limit adventure playgrounds' liability would be necessary to make room for the reasonable and unavoidable risk taking inherent in unstructured play. European playground safety standards, for instance, exempt adventure playgrounds from meeting their requirements.[22] The need for adventure playgrounds to have staff suggests that financial barriers may be part of the reason that only a handful of adventure playgrounds exist in North America. Partisans for thick communities could look to fill the financial gap by seeking to tax privatized playgrounds in order to subsidize public alternatives. Targets might include the private indoor playgrounds whose profits partly depend on "stranger danger" worries, charging parents upward of seven dollars per hour per child for access to a sterile but completely secure play area.

The development of adventure playgrounds would likely have fairly limited effects without broader changes. Parental fears stem in part from their own poor integration into their local social environs. The fact that the idea of their children encountering fellow residents or strangers on city streets or in parks elicits such anxiety in many parents can be attributed to the fact that they have few established relationships with neighbors who can be their proxy eyes and ears. Parents face what philosopher Robert Kirkman has referred to as a "limits of agency."[23] Even if they wanted to raise their children in more communitarian ways, the barriers posed by the design and sociocultural makeup of many neighborhoods make doing so infeasible, or doable only at the sacrifice of other important values (e.g., perceived safety). At a minimum, time is necessary for communitarian interventions to begin to lessen the culture of aloofness that stands in the way. Carving out more room for relationally thick and geographically proximate spaces for youth to exercise their autonomy could gradually open other avenues for thickening communal relationships.

At the same time, there are certain risks to avoid when pursuing communitarian technologies like adventure playgrounds, namely nostalgia. For instance, free-play advocate Mike Lanza's heart has been in the right place, from a communitarian perspective, when he has lamented the demise of neighborhoods that permitted children to collaboratively structure their own activities and build local social bonds; however, he has pinned the blame for such changes on the rise of "mom philosophy," claiming that the term *bully* pathologizes the previously "normal" aggression of boys.[24] As laudable as Lanza's work promoting free play is, the gendered language through which he understands childhood leaves a lot to be desired. Given the growing recognition that twentieth-century ideas about "normal" masculinity have been used to justify sexual assault, discriminate against gay men, and constrain the emotional development of boys,[25] communitarians would be better off not tying their advocacy of free play to idealizations of 1960s manhood. Such language inevitably excludes girls and demeans traditionally feminine forms of free play. One can yearn for and promote more free-range childhoods without reproducing the cultural ills of previous, more free-range generations.

Opposing Technologies That Privatize or Displace Thick Community

Developing more thick communitarian societies would entail ensuring that people spend less time watching television or surfing the Internet in their air-conditioned dens and more time conversing on their front porches and hanging out in public places. Such practices, however, would likely be hardly affected by directly penalizing citizens for their television and air conditioner usage. Given typical deficits in quality public spaces and the nonexistence of front porches, gathering places, and even sidewalks in many North American neighborhoods, citizens cocoon because they have few other feasible options available to them. Hence, alterations to urban form, among other changes, would be necessary to nudge them toward public sociability.

Because television and other forms of mass media are likely to continue to dominate the everyday lives of citizens in affluent nations, short-term goals would probably need to be modest. One incremental change would be to encourage residents to do more of their TV viewing in public. In Europe, even the smallest towns will erect public viewing screens for popular soccer matches. Doing the same more often for sports in North America would get a greater portion of the population out of their homes. The staging of free public viewings at community theaters would be the best option where the weather is often prohibitively cold or inclement. Many municipalities, however, may not have the budget for such events. Franchise law, which is written to give cities the negotiating power to ensure that the provision of cable television to residents meets "community needs,"[26] might be productively leveraged in such cases. Using the argument that cable television in private homes is in itself a direct threat to their community needs, municipalities could require that cable companies foot part of the bill for public viewing. Some cable companies might object that they would, in turn, lose some portion of current subscribers. Yet, on the other hand, they would likely be broadening the reach of their advertising and content. Cable companies would therefore have some grounds for negotiating larger payments from advertisers or changes to the license fees paid to content providers to make up for any decreased profits.

A similar approach could help make an escape from the summer heat more of a public good than a private one. Summer air conditioning uses up valuable nonrenewable resources, contributing to ecological harms such as climate change. According to some estimates, air conditioning constitutes up to one-third of peak summer energy usage, though this figure no doubt varies according to geography and culture.[27] Europeans, for instance, are notably less reliant on air conditioning than Americans. Increased summer demand for A/C, moreover, requires the construction of larger or more numerous power plants as well as distribution system upgrades. Hence, there is already a strong environmental reason to financially penalize summer air-conditioning use via *dynamic peak pricing* of electricity, which charges higher rates at peak use hours, among other schemes.[28] Advocates would better maximize the communitarian gain of such measures by ensuring that some portion of peak pricing revenues support the building of public cooling technologies, namely fountains and heavily shaded parks. Additionally, peak pricing penalties could be tiered, penalizing private homes and regional malls the most but community centers and "third places" the least.

What about cities and towns with mild summers and frigid winters? How could warmth be more of a public good? Exposure to the cold is often seen as a greater danger to the health and well-being of humans than warm temperatures, apart from heat waves. Increasing charges on electricity or gas used for home heating is likely to be both politically infeasible and opposed on humanitarian grounds. At the same time, few cities publicly provide outdoor warmth. As architectural scholar Norman Pressman noted, "[Urban] designs and realizations are similar whether in Oslo or Miami, Toronto or Phoenix, Reykjavik or Los Angeles."[29] The usual answer to cold weather is to construct expensive skywalks or underground passageways, connecting malls with other downtown buildings. The trouble with these systems is that, by segregating retail services for middle- to upper-class downtown workers and shoppers to private walkways, they contribute to a depopulating of outdoor sidewalks and undermining of street-level business.[30] At best, their hermetically sealed walls are attractive for only a few months in any given year.

The "solution" provided by skywalks, underground shopping districts, and giant malls is a brute force response: an attempt to erase geographic and seasonal variance rather than working with it. The reasons for this are not really technical but political and cultural. Indeed, architectural research does exist concerning how to enhance the thermal livability of winter cities,[31] which shows ways to supply public heating without the use of fossil fuels or expensive HVAC systems. Many attendees of wintertime sporting matches in open stadiums, for instance, have found themselves unzipping their jackets and loosening their scarves after an hour or two; shielding from the wind combined with the heat produced by human bodies can quickly create a surprisingly pleasant microclimate. Similar principles can be put to work on streets, sidewalks, and in winter parks by sheltering them from cold northerly (or southerly) winds. Moreover, large sun reflectors and required southern (or northern) orientation could produce bright and warm wintertime public spaces. The largest barrier to technologies of public warmth, therefore, is not technical or financial. Rather, ignorance and the absence of supportive regulation on the part of city planners and urban designers stymies their implementation.[32] Public professionals need the proper training and the regulatory impetus to more often incorporate features such as wind breaks, solar orientation, thermal sinks, and outdoor arcades into public spaces.

Even with public spaces whose thermal properties might better entice residents out of their dens and family rooms, cell phones and other digital devices too frequently help produce outdoor areas uncongenial to public sociability. How might their usage be better regulated? One possibility would be to allow and encourage parks, pubs, and cafés to restrict cell phone use and Wi-Fi signals. In the same way that many restaurants used to have nonsmoking and smoking areas, locales might have separate areas for Wi-Fi and cell phone use. Public spaces could have designated booths for making cell phone calls instead of permitting them everywhere.

Rather than rely on workers to police the use of devices, such measures would have a more substantive effect if Wi-Fi and cell signal were electronically jammed in unapproved spaces. Municipalities and businesses in the United States, however, are prevented from experimenting with such systems by federal law. It is illegal to import, market, sell, or use any device that interferes with wireless signals. The reasoning provided for such regulation appears entirely based on worries about potential interference with calls to emergency services. France and Japan, in contrast, have permitted cell phone jamming devices under the condition that they do not affect emergency communication. Indeed, there is no inherent technical reason that jamming devices in public spaces could not be designed to allow emergency calls and paging functions for particularly pressing communication.[33] Another barrier to changing regulations is the interpretation of access to the wireless communication spectrum as private property, which construes signal blocking as equivalent to stealing. This legal interpretation leaves little room for the consideration of the trade-offs between the private use of digital devices and public sociability as a collective good.

Finally, a greater amount of research and innovation could be directed toward developing more prosocial apps and devices. Such technologies would work by lessening the psychocognitive barriers to public sociability. Consider a device called the Coffee Connector. Meant to encourage mingling, it dispenses coffee only when two people both enter their names and interests into its screen.[34] Similarly, apps such as Tinder already alert users to the existence of proximate others with similar interests, although they are much more directed at facilitating dating and semianonymous sex than social community. Nevertheless, one could imagine coffee shops and other hangouts having on-premise apps that include a public chat function where users could signal their interests and openness to conversation. Such apps, given that they are unlikely to provide huge financial returns, would probably need public subsidy to be written. Alternatively, pubs could have solo patrons input such information into a tablet and suggest possible discussion partners. Though somewhat artificial and contrived, such measures might be a big help for citizens who have become overly habituated to social privatism in public spaces.

Discouraging Technological Retreats from Community

Some devices may be almost wholly incompatible with the practice of thick community. Concealed-carry handguns represent a withdrawal from the possibility of collectively achieving security, and the provision of companion robots signifies a collective acknowledgment of defeat concerning finding human relationships for the chronically lonely. What could be done about the technologies that are being advanced to fill the vacuum left by declines in some of the seven dimensions of thick community?

Like other personal technologies, targeting concealed-carry handguns or companion robots directly would likely be the least efficacious strategy. The continued entrenchment of narrowly individualistic political worldviews makes it increasingly difficult to intervene in technological matters that have already been framed in terms of personal choice. Recent attempts to employ gun-control legislation in the United States have faltered in the face of heavy opposition by the National Rifle Association and firearm companies, who argue that citizens' individual rights to personal security via firearms automatically trump all other concerns. Moreover, opponents to stricter regulations of companion robots would likely claim that restrictions would deprive the chronically lonely of choice or would "legislate morality." The widespread tendency to view technologies as mere volition enhancers and ignore their political biases only further entrenches narrowly choice-oriented thinking on technological governance. Given this broader overarching barrier, the most productive avenue toward lessening the damage these technologies do to thick community would be to focus on developing communitarian alternatives: community policing, volunteer public houses, and better residential options for retirees.

Making policing more communitarian faces several barriers. Some of the challenges are budgetary. Similar to a Wal-Mart or a leisure center, highly centralized policing arrangements are much less expensive, allowing underfunded departments to cover more territory with fewer officers. With budgets already set for centralized policing, a transition to community policing often means officers being assigned to "beats" much too large for effective community engagement. The nonwalkability of diffuse suburban environments, moreover, keeps officers isolated within their cars as much as it does for average citizens. The sociotechnical momentum of suburbia thereby helps ensure the continued existence of highly centralized, insular police forces.

Cultural and organizational momentum create additional hurdles. The successful implementation of community policing by the New York Housing Authority from the 1950s until the 1970s was partly a result of officers being held accountable to civilian project managers as well as encouraged or enforced to meet with citizens.[35] In fact, their offices were located in record rooms within project apartment buildings, rather than at a central precinct. Studies find, however, that traditional police department officials are highly resistant to the decentralizing of structure, the civilian oversight, and the public interactions entailed by community policing.[36] In contrast, many departments appear to merely rhetorically embrace the philosophy of community policing; the structural changes needed to make it work are seldom made.

In fact, studies of officers, at least in the United States, frequently find that they denigrate community engagement and collaboration as not "real" police work.[37] One potential cause of this attitude is the media-reinforced cultural archetype of the "warrior" cop getting in shootouts and arresting bad guys. It is further encouraged to the extent to which policing institutions incentivize arrests and fail to reward community policing tasks. As long as dominant police culture frames small talk with residents or attending community events as less worthy tasks, the embrace of community policing is unlikely to be substantial. The dominant frame of cop-as-warrior comes at the expense of cop-as-community-member. In any case, community policing is often denigrated by officers because they lack good community relations to start with, especially in extremely large, centralized police forces. A vicious circle persists wherein officers may turn to traditional tough guy cop culture in order to feel a sense of belonging, because overly large "beats," squad car cocooning, and a lack of public oversight prevents them from achieving a feeling of togetherness with the citizens they ostensibly serve and protect.

Finally, there is a legitimate concern regarding *which community* community policing serves. Without ensuring the inclusion of marginal populations, such as youth and poor minorities, and the existence of concrete means of shaping policing practices, community policing can end up serving the interests of already advantaged community members. It risks amounting to policing *through community* rather than *by it*.[38]

Mitigating these barriers is no easy task. Police unions, as evidenced by their obstinacy in the face of recent protests over police brutality, are frequently political opponents of any measure that threatens the autonomy of policing institutions from civilian oversight. The most feasible strategy, given the obduracy of established policing sociotechnical systems, might be for municipalities to emulate the creation of New York City's Housing Authority and develop parallel, civilian-managed policing institutions.[39] Moreover, they could do so under the auspices of "freeing" traditional departments from walking "beats" or doing patrols, tasks that officers steeped in warrior cop culture are likely to think of as beneath them anyway.

Alternative institutions, however, are only likely to develop with substantial governmental support. The increasing visibility of incidents of excessive force, such as the shooting of John Crawford by police for holding a toy BB gun inside a Wal-Mart or an officer gunning down a teenager in Ferguson, Missouri, could be leveraged to open up federal funding for alternative policing institutions. The poor minority neighborhoods where such alternatives would be desirable, however, are often already served by regional police departments unlikely to appreciate the competition. In large metropolitan areas, the racial and class-based fears of suburbanites frequently translates into unwavering support for antagonistic, arrest-focused policing. If these barriers prove to be too difficult to overcome, state- or national-level policy would need to step in to permit neighborhoods suffering from officer-perpetrated human rights abuses to secede from their policing district and develop their own alternative institutions.

At least some of the suburbanites who turn to personal firearms or promote antagonistic policing to realize a sense of security are motivated by the fear of racially different others. They would be unlikely to buy into community policing while those fears persist. Antagonistic and disproportionate policing of racial minorities forms a vicious circle with the perception that they are dangerous, justifying further fear-based targeting. Such perceptions may not be feasibly lessened until poor minority neighborhoods are able to substantively implement community policing, breaking the cycle.

In the same way that discouraging personal handguns demands a reorganization of the provision of policing and security, stemming the potential tide of companion robots would require devising organizational solutions to the problem of loneliness. One of the main markets envisioned for companion robots is elderly citizens who either live alone or in institutionalized "old folks" homes. For the former, their children, if they have any, often live far away or are too busy to see them often. Furthermore, many of their close friends and other ties may have already passed away. Similarly, institutionalized seniors are frequently isolated, sometimes receiving few or no visitors and unlikely to get much social intimacy from overworked and underpaid staff. Indeed, studies typically find that nursing home residents spend the majority of their days alone in their rooms, doing little or nothing with their time.[40]

One solution for the problem of lonely seniors, at least for those with some mobility left, would be to reform urban environments to include more public spaces, to be more walkable, and to contain a larger number of well-supported third places. Such changes would no doubt take years, more likely decades, to implement. In the interim, older people are typically presented with two options: volunteering and senior living arrangements.

Contemporary seniors volunteer more than other age groups but suffer from loneliness at relatively higher rates.[41] Such rates suggest that at least some of them either do not have good opportunities to volunteer or that barriers stand in the way of their participation. Indeed, studies find that factors like ageism, logistical difficulties, lack of confidence, and communication troubles discourage volunteering among seniors.[42] Potential elderly volunteers are dispirited by cultural presumptions that old age implies lack of competency as much as by transportation difficulties and the expectation that they communicate via more recently developed media devices. What retired seniors may need most is a hangout and third place that has the second purpose of facilitating volunteering: a *public house* in the full sense of the term. One could imagine a publicly supported café or pub where a volunteer ombudsperson keeps office hours and connects retirees with those requesting help, perhaps taking responsibility to relay email and social networking messages in print or in person. Maintaining a physical location would be particularly apt given that those most receptive to volunteering are the recently retired.[43] Such persons are already habituated to going somewhere for work every day and could easily develop the daily habit of visiting the volunteer public house, especially if they miss the embodied communality and small talk previously provided by their jobs.

Another popular option for lonely seniors is to move into dormitory-like retirement or independent living centers, which provide studios or one- to two-bedroom apartments alongside game rooms, cafeterias, and other amenities. This option, however, is only available to the fairly affluent. The typical cost of a one-bedroom apartment in centers run by Holiday Retirement, for instance, is around $2,500 per month, which exceeds the median wage in the United States and, therefore, probably most people's retirement budgets.[44]

An alternative solution would be to induce adult children to more often house their aging parents. The barriers to this strategy are primarily legal and cultural. Municipalities and homeowners' associations in the United States often ban or exert considerable regulatory pressure against *granny flats* or *secondary suites*: small apartments constructed over garages or in the basements of detached homes. The practice is allowed in no more than a few select neighborhoods in Minneapolis, and San Francisco only started permitting their use and construction a few years ago.[45] Such laws, apart from greatly reducing the supply of affordable housing in suburban neighborhoods, deprive citizens of the option to house their elderly parents in ways that maintain the level of privacy to which they are accustomed. Banning secondary suites removes the middle option between living completely separately and becoming housemates. Eliminating such bans, however, may not do enough. Low-interest loans might be necessary to spur their development. One illustration of this is how the Canada Mortgage and Housing Corporation offers forgivable loans for the building of secondary suites for low-income seniors and the disabled on First Nations reserves.[46]

Although the prevalence of multigenerational homes is on the rise,[47] they are far from common. Some of the resistance to secondary suites or more extended family households stems from decades of cultural changes that have led to the idealization of small households, whether composed of nuclear families, roommates, couples, or solo individuals. Similar to the forces that have motivated solo dwelling, the perceived infeasibility of multigenerational homes likely stems from a desire to avoid conflict. Barring physical barriers, it is much easier, though not financially, to live in one's own home than to negotiate new arrangements of authority and responsibility with one's aging parents. Certainly some members of extended households have, at one time or another, felt oppressed by an elderly patriarch or overly patronized by their own children. Giving up on extended families because of such occurrences, however, is like abandoning marriage and family life because many people have not yet learned to be good parents and spouses. Nevertheless, ending the vicious cycle between familial fragmentation and weak conflict negotiation skills is not something done by housing policy. Indeed, it would take some combination of changes, from democratic schooling and community-owned infrastructure to adventure playgrounds and more communal domestic environments, for a greater number of citizens to find multigenerational living both desirable and feasible.

Institutionalized seniors, on the other hand, could more often have their needs for social contact and belonging met if daycares and nursing homes were colocated. Given that both services suffer from ever increasingly tight budgets, colocation can save costs at the same time that it leverages seniors' needs for meaningful social contact into caring for and educating young people. Such an arrangement would even help develop more communitarian adults. Indeed, one study of a shared-site child and adult daycare center found that intergenerational play increased children's social acceptance of and empathy for the elderly. The benefits to the participating seniors were similarly immeasurable. One summarized her sentiments saying, "The children need to be here ... I do not feel sad here."[48] Moreover, given all the research demonstrating the significant health advantages resulting from social belonging and community,[49] nursing home residents who interact with children would likely require far fewer medical interventions. Indeed, colocating nursing homes and daycares might be considered a form of preventive care.

Despite their clear advantages for social well-being and potential to lower the costs of senior and child care, colocated facilities remain rare. In certain areas, the barriers are legal. In fact, California law prohibits placing daycare and senior care in the same building.[50] Some centers, however, have been able to skirt this rule by constructing two separate buildings connected by a breezeway. Other significant barriers stem from entrenched patterns of thought and ignorance. The dominant structure of public schooling in most technological societies frames same-age peer groups as the norm.[51] Having spent decades cultivating friendships only with those within a few years of themselves, many citizens are unlikely to think of intergenerational social ties as attractive or perhaps even possible. Hence, it makes sense that state legislators would unthinkingly prohibit intergenerational care facilities and that those designing care centers would only seldom consider mixing age groups. It may take bold action by state and national governments to shift thinking about daycares and nursing homes. If even a mere fraction of the millions of public R&D dollars going to companion robotics research were redirected toward experimenting with and establishing intergenerational care facilities, such approaches could garner greater recognition and legitimacy.

Promoting Communitarian Repair Cultures

Finally, artifacts could be more compatible with communities of repair and tinkering. Consider the vibrant, albeit networked, communities of practice forming around open source and free software. Similarly, *maker spaces* have gained considerable popularity. In such spaces, hobbyists and budding tech entrepreneurs tinker and design, often simultaneously hosting DIY Bio meetings where enthusiasts play around with DNA and biological organisms. Limiting the potential of these groups is the fact that they tend to require a level of expertise, income, and time out of reach for the average person. Indeed, designer/activist Aral Balkan refers to open source software as a "sandbox for enthusiasts," going so far as to argue that the idea that the open source movement can empower the average consumer is based on a misguided "trickle-down" theory of technology.[52] It presumes, without evidence, that expert enthusiasts building technology for other enthusiasts will somehow trickle down to produce substantial benefits for laypeople. Maker spaces and DIY Bio labs seem hardly different. Skimming their webpages one typically finds membership rolls composed of people with college degrees in science or engineering and occasionally fine arts. Finally, many, if not most, of these spaces charge upward of a hundred dollars per month for a membership. How could a greater diversity of people have the opportunity to work alongside neighbors and others in repairing, tinkering, and building?

Tool libraries could be part of the solution, though their current reach remains limited. As one review of Oregon's tool libraries notes, they are typically open a couple days each week and are not publicly funded.[53] A straightforward fix would be to more significantly integrate tool libraries into the public library system. Given that the existence of contemporary information networks somewhat obviates the need for physical books, libraries would better capitalize on their strengths by focusing on providing public access to material things that cannot be easily digitized. Doing so would enable tool libraries to serve a wider range of communities as well as open up possibilities for funding full-time tool library workers and providing free classes. This latter possibility is necessary because the barrier to more citizens working together to build, repair, and maintain their technologies is not merely a lack of tools, but access to the kind of experience that one cannot get from a YouTube video. Publicly funded, full-time tool librarians could shift the Fixer movement from a hobbyist group of affluent whites into an everyday practice across races and classes.

As helpful as such changes might be, however, these efforts do not address a core barrier: technologies that are difficult or expensive to repair and modify. The fact that companies such as Apple face no repercussions for using nonstandard screws or epoxy to seal their devices shut lets firms technologically prohibit people from engaging with the inner workings of their gadgets. It would probably take sizable incentives or tax penalties to convince firms to make their products more easily repairable, given that it undermines their revenues from repair services and sales.

Nevertheless, they might be more indirectly incentivized to do so. So-called take back laws require companies to recover their products from customers when no longer wanted or usable, thus placing the responsibility on firms to ensure proper disposal of consumer goods.[54] Apart from the clear environmental benefits resulting from companies being encouraged to design products so that they are easier to remanufacture, recycle, or dispose of safely, such laws could result in devices that are more straightforward for users to disassemble and modify. As economics analysts have pointed out, however, take back laws only do this when other firms are not allowed to intervene and compete for broken or obsolete products.[55] For instance, a company that makes money by purchasing old electronic devices directly from consumers and melting them down for their valuable heavy metals would enjoy all the increased profits resulting from easier to disassemble devices while leaving the original manufacturer to cover the increased costs. Such an arrangement would result in original manufacturers having a much weaker incentive toward improved recyclability and reparability, doing less to discourage consumers from parting with their still repairable technologies.

Wherever dominant firms cannot be easily steered or incentivized, governments could redirect public research and development funds toward designing more easily repairable technologies. Indeed, technology policy analysts Edward Woodhouse and Daniel Sarewitz suggested this strategy as a way to ensure that R&D more often serves the needs of the poor.[56] Going one step further would be to require that resulting designs be placed in the public domain. This would contrast the current paradigm of public research funding, wherein public dollars are used to develop products that private companies quickly patent and restrict to maximize their profits. In any case, the resulting knowledge and designs could be handed over to new kinds of firms, perhaps even cooperatives, whose focus would be on making a small but reasonable profit producing reliable, affordable, and open technologies.

Lessening the community thinning effects of many techniques and gadgets would entail not only directly discouraging their use but supporting more vibrant communitarian alternatives. Though TV licensing fees and dynamic peak pricing would be helpful, public viewings and public technologies for warmth and cooling may be more effective at peeling users away from their televisions and air-conditioning units. If parents are to more often experiment with cosleeping and other alternatives to crying it out, they would need more flexible work arrangements—not just less culturally biased pediatric advice.

Efforts to alter contemporary techniques and artifacts face broader and more tenacious cultural barriers than those to reform organizations or infrastructures. Techniques can persist long after their more formal or institutionalized support networks have disappeared. Even if the American Academy of Pediatrics were to reverse its position on cosleeping, ideas about independence and risk would continue for years, if not decades. Moreover, many members of technological societies are likely to continue thinking of techniques and technologies as matters of personal choice, especially in highly individualistic countries like the United States. *Technological liberalism* will probably remain a dominant political-cultural ideology for the foreseeable future.[57] The idea that technology writ large can provide a neutral foundation from which individuals pursue their own ideas of the good life will continue blinding many citizens to how technological systems and practices render some ways of living more difficult than others. A person looking to enjoy thick community has few options and little agency as an individual. In contrast, buying a smartphone or registering for a Facebook account is dead easy. Insofar as technological liberalism prevents more citizens from recognizing how their lives are subtly constrained by their sociotechnical contexts, as if by legislation, their ability to collectively do anything about it will remain highly restrained.

***

The purpose of chapters 7–9 has been to illustrate how status quo technological arrangements are nothing natural but are the product of a constellation of cultural, regulatory, economic, and other social factors. The fact that many members of technological civilization currently live in diffuse suburbs, keep their children sequestered in their homes, meet their material needs through individuated exchanges on retail and infrastructural networks, and turn to private alternatives to community engagement is not really the result of anything resembling free individual choice. People's decisions are shaped to a significant extent by sociotechnical forces lying largely outside their control and frequently unbeknownst to them. A vast array of public subsidies and invisible support ensure the momentum of suburbia, regional malls, automobility, and the oligopoly of giant retail corporations, including overlooked sources like the near mandatory provision of free parking, inadequate labor laws, and the failure to incorporate commuting and utility costs in mortgage calculations. In the language of STS, these technologies—as well as their momentum and obduracy—have been socially constructed.

I have made several recommendations for lessening, leveraging, or skirting these barriers. It is important to keep in mind that no single intervention into the sociotechnical systems shaping everyday life is likely to significantly foster thick community on its own. Just as people did not become networked individuals immediately after incorporating a few networked technologies into their lives, but did so rather over the course of decades of incremental but widespread sociotechnical changes, walkable and mixed-use neighborhoods or community-scale energy systems are unlikely to produce thick communitarian societies on their own or very quickly. Community is a vector phenomenon, a product of a constellation of factors. Any single change is best framed as a small nudge toward thick community or perhaps even merely a catalyst for other nudges. Like a chemical reaction, various sociotechnical ingredients of thick community probably need to be present in sufficient amounts, along with enough added energy, for social evolution toward more communitarian ways of life to occur. The expectation that altering a single variable should have dramatic effects can itself be a barrier to improved thought and action. It works under the misguided assumption that social life is a simple, linear, and almost mechanistic phenomenon, rather than intricate, nonlinear, and probabilistic.

Broader economic factors likewise stymie the development of alternative and more communitarian technologies. Automobility, sprawling suburbs, distant retailing, large-scale energy infrastructures, and other technologies of networked individualism have emerged and persisted in an age characterized by a glut of inexpensive fossil fuels and other nonrenewable resources, and distinguished by widespread toleration of ecological disruption. When the environmental scholars who think and write incisively on the ever-pressing need for more sustainable, no-growth economies describe more desirable futures, they typically contain many of the communitarian features that I have discussed.[58] The processes that prevent social evolution toward more sustainable ways of living likely hinder movement toward thick communitarian societies as well.

Although economic localism and alternatives to consumerism are greatly more efficient from a resource standpoint, they are less likely to be realized insofar as broader economic policy fails to take growing resource scarcity and potentially catastrophic changes to climate and environment seriously enough. Although incremental steps toward thicker communities could no doubt be made in the short term, more substantive changes may require shifting away from growth-oriented economies. Apart from the numerous barriers erected by those who currently benefit from an economic paradigm rooted in a belief in endless growth, at least some citizens would be likely to equate the end of consumer society with the demise of civilization itself. Fewer people have as much experience with economic dependence on friends, family, and neighbors as in previous decades and centuries. Many would probably find the prospect of losing the appearance of self-sufficiency enabled by large-scale consumer markets to be frightening. Nevertheless, the central claim of groups like the Transition Town Movement, who aim for a smoother transition to a post-peak oil and climate change world, has been that it is far better to adapt sociotechnical systems to that potential future now, while some major resources can still be mobilized to transitioning, rather than wait until technological societies are teetering at the edge of collapse.

Regardless, as helpful as my illustration of barriers and potential solutions might be to advocates of thick community, any set of proposals is likely to be incomplete in the face of the uncertainties and complexities of reality. What is most needed, if citizens are to realize thick communitarian technological societies, is an enhanced ability to learn from sociotechnical change, prevent foreseeable harms to communal ways of life, and more broadly pursue the intelligent and democratic steering of technological innovation.

Despite all the risks and undesirable unintended consequences produced by technological innovations,[1] rarely have they been consciously and intelligently governed to lessen or more quickly respond to deleterious effects. In contrast to discourse that frames technological change as a natural evolutionary process, a more apt metaphor is that of a driver steering an automobile, who has the option of steering, accelerating, and braking in response to where he or she wants to go. Often it seems like innovation proceeds, as Neil Postman and Charles Weingartner described, as if important decision makers were "driving a multimillion dollar sports car, screaming, 'Faster! Faster!' while peering fixedly into the rearview mirror."[2] Although the two authors were describing the state of the mid-twentieth-century education industry rather than technological innovation writ large, the image remains apt. Massive technological undertakings such as nuclear energy and the space shuttle, as political analysts have shown, largely proceeded with the assumption that the relevant experts had it all figured out, only to be proved wrong by catastrophic or near-catastrophic accidents and products that came in over budget, past due, and nowhere close to delivering on early promises.[3] Today, one thinks of Toyota's attempts to ignore sticky electronic accelerator pedals, which subjected some drivers (and their passengers) to uncontrolled accelerations, or British Petroleum's negligence in the blowout of the Deepwater Horizon oil rig, which gushed some five million barrels of oil into the Gulf of Mexico over eighty-seven days in 2010.

Much the same tends to be true of technological innovation with respect to what could be termed *psychocultural goods* like community, although these goods are more seldom considered by those studying the consequences of emerging technologies. As Edward Woodhouse noted, Philo T. Farnsworth could hardly foresee the broader cultural changes that would result from his invention of television tubes, namely their role in civic decline as they became omnipresent in North American homes.[4]

The failure to adequately govern sociotechnical change partly stems from the belief that technological development evolves autonomously, "progressing" societies all on its own. Consider historian Daniel Boorstin's assertion that "the advance of technology brings nations together" with "crushing inevitability" and his celebration of TV for "its power to disband armies, to cashier presidents, to create a whole new democratic world."[5] Boorstin's attitude no doubt reflected the broader reluctance of mid-twentieth-century citizens to seriously consider the potential of technologies to enable undesirable or unwanted changes. Similar sentiments were common a century ago regarding electrification and have remained prevalent—consider the contemporary belief that the Internet will automatically democratize despotic governments.[6]

Although considerable advances in thickening different dimensions of community life would be enabled by addressing the sociopolitical barriers outlined in the last three chapters, any gains would likely be partial and short lived without more foresighted public decision making regarding technological innovation. How would new innovations need to be governed given the fact that "as we invent new technical systems, we also invent the kinds of people who will use them and be affected by them?"[7] How exactly might citizens more intelligently steer technological development in ways that protect, sustain, or enhance the experience and practice of thick community? How could intelligent trial-and-error strategies be applied to technological innovation to protect against the potential risks to community life?

Intelligent trial and error, however, is as much a political framework for avoiding unintended consequences as a set of strategies for assuring organizational success. How could trial-and-error strategies help improve the implementation of communitarian technologies? Finally, how can the framework be extended to already built and obdurate technologies? Indeed, many of the technologies that I have described thus far, such as suburbia and limited-access freeways, are massive sociotechnical barriers to thicker community life. Communitarians would not only benefit from the more intelligent steering of new innovations but also from the ability to prudently dismantle or reconstruct existing technologies.

Intelligent Trial and Error

The intelligent trial-and-error (ITE) framework has been developed by political scientists engaged in the study of risky technologies and organizational mistakes.[8] ITE begins with a call for early deliberation by well-informed and diverse participants. Decision-making processes would, in turn, fairly represent those who might be affected by an innovation, be highly transparent, and broadly distribute both the burden of proof and authority to decide. If deliberations result in the choice to proceed, innovation would continue prudently and with adequate preparations to learn from experience. That is, activities including premarket testing, establishing redundant back-up systems, building in flexibility, and scaling up gradually would help to ensure that mistakes would be either averted or as small as possible. Moreover, the existence of diverse, well-funded monitoring groups and disincentives for mistakes, such as fees and fines establishing victim funds, would create a policy environment that encourages error correction. Finally, deliberation, testing, and monitoring would involve relevant experts, and their results would be broadly communicated. Advisory assistance, especially concerning the environmental and social effects, would also be readily available to have-nots, ensuring equity. The above reforms, furthermore, would be unlikely to be effective unless polities instituted significant protections against widely shared conflicts among technical experts, business executives, politicians, and other elites.

Some degree of trial-and-error learning no doubt occurs for most technological innovations; the problem is that learning frequently happens too late for easy error correction. Consider the Internet. A lack of ITE during its development has harmed cyberasocials—people who are unable or unwilling to feel a sense of social copresence or belonging through digital devices—and others whose needs for thick community are poorly served by contemporary communication networks. Any political deliberations regarding the psychocultural risks of the emerging Internet appear to have been largely ineffective, given that today's Web is ambivalent at best with regards to thick community. This ineffectiveness is further clear from the absence of direct steering of the Internet's development by governments with respect to social values and goals, despite substantial humanistic concerns being raised throughout its history.[9] The relative lack of controversy surrounding governmental decisions, such as the one to shift from the research-focused NSFNET to the commercial Internet and the choice to subsidize high-speed Internet infrastructures, gives the impression that such moves were made without any conscious consideration of the potential unintended consequences.

Much like other emerging and similarly hyped technologies, diverse and divergent partisan positions regarding the Internet were frequently dismissed. As journalist Lee Siegel wrote in 2008, "Anyone who does challenge Internet shibboleths gets called fuddy-duddy or reactionary."[10] A

Table 10.1

Major Components and Requirements of Intelligent Trial and Error

Major Components of ITE	Requirements
Deliberation	- Started early
	- Fair representation of those potentially affected
	- Transparent
	- Burden of proof and decision authority shared
Precaution	- Initial testing
	- Gradual scale-up
	- Redundant safety measures
	- Built-in flexibility
Error reduction	- Recognition of need to learn
	- Well-funded, diverse monitoring groups
	- Disincentives for resistance to error correction
	- Incentives for error correction
Analysis and communication	- Sophisticated social and environmental analysis
	- Protections against conflicts of interest
	- Findings readily available through appropriate means
	- Extra assistance to have-not participants

debate framed much in the same way the former U.S. president George W. Bush presented the second Iraq War, "You're either for [the Internet] or against [the Internet]," was woefully unprepared to head off the unintended consequences. Most societies jumped headlong into the process of restructuring their informational and communicative infrastructures vis-à-vis the Internet.

Active preparation for learning about the potential social effects of digital mediation would have entailed a very different evolution of the Internet. Experimental testing uncovering the existence of cyberasocials could have been performed in 1994 rather than in 2014.[11] Doing so would not only have tempered hyperbolic rhetoric about the coming "global village" but also could have identified and perhaps even energized a clear set of stakeholders likely to be harmed by the increasing digital mediation of social life. Cyberasocials could have then worked alongside advocates for local, embodied community, and against Silicon Valley cyberutopians, in deliberations regarding how to best proceed with computing technologies. Rather than pursue a poorly controlled, market-led diffusion of Internet technologies, with psychological and sociological research being performed largely too late to influence decision making, deployment could have proceeded more gradually. Governments might have pushed for certain cities and neighborhoods to be wired earlier than others and funded dozens of longitudinal studies similar to that performed in a Toronto suburb (Netville) during the 1990s.[12] Such studies, moreover, might have had an explicit focus on the consequences for cyberasocials and involved staunch critics, not just apologists. Results from these studies could have then been funneled back into venues for deliberation.

If the decision had been made to proceed, funds could have been set aside to compensate the people harmed if and when existing modes of community life became partly undermined by digital technologies. Those unable to feel a close social connection through online social networks would have benefited from a tax on Internet access and digital devices that, in turn, subsidized more communitarian neighborhood design, publicly oriented pubs and cafés, and technologies that could have helped them to limit the intrusion of Internet technologies into their social lives. Internet firms might have been incentivized to experiment with ways of providing Internet access that coincided better with the geographic reach of local community. Internet service providers could have been encouraged to set up networks that afforded substantially higher local data speeds and special applications that work only within neighborhoods and districts, incentivizing the use of such technologies to connect with more proximate friends and loved ones.

ITE, Community, and Emerging Technologies

Looking back on the past is only helpful, however, for illustrating how the strategies that make up intelligent trial and error could have been useful. The more important matter is the application of ITE to emerging technologies. How might research and development of driverless cars and companion robots be steered to protect against potential undesired effects on community?

Driverless Cars, Sprawl, and Job Loss   As is the case for most hyped innovations, public debate on driverless cars in various media outlets rarely includes a careful consideration of the full spectrum of potential risks. Some observers, such as a blogger for *Freakanomics*, breathlessly describe the autonomously driving automobile as a "miracle innovation" and a "pending revolution." Even the more circumspect take by ethicist Patrick Lin was focused fairly narrowly on the ethical issues arising in the process of *adaptation* rather than *conscious governing*, namely how to assign blame when accidents occur, the possible effects on the insurance industry, and the risks of hacked driverless cars. Moreover, Lin ended his analysis with unmitigated technological determinism: "The technology is coming either way.

Change is inescapable." Authors of a Mercatus Center white paper likewise contended that "social progress" will be best served by opening up space for driverless cars to be an area of "permissionless innovation."[13] They argued that governments need to limit regulation of autonomous automobiles so that firms can "innovate" with abandon, working under the presumption that everything always works out for the best when innovators proceed largely unhindered.

Hence, the first major barrier to more prudently steering driverless car technology is the lack of recognition of the need for ITE learning. Despite some forty years of scholarly activity within science and technology studies, many people continue to believe that technological development proceeds autonomously, that technical change automatically results in humanistic advancement, and that everyone benefits equally from technological innovation; otherwise intelligent journalists, politicians, engineers, and professional ethicists have not stopped embracing incorrect ideas about technological change. Getting over this initial hurdle will be necessary before ITE-like governance could be feasibly applied to driverless cars or any other emerging technology.

The self-driving car is still, nevertheless, in its incipient stages, the ideal time for effective deliberation regarding the potential consequences. Although Google technicians have logged thousands of hours of test driving, their cars have not yet been fully deployed—although forms of automated assistance have recently become features in high-end automobiles. Besides the harms accruing to car salesmen, cab drivers, and long-distance truckers potentially put out of work, there are real concerns about potential secondary and tertiary effects of autonomous vehicles on urban landscapes. Positive unintended consequences are, of course, possible. If deployed as more inexpensive cabs, decreased rates of private car ownership might lead to denser urban environments—given the lessened need for copious amounts of parking. On the other hand, self-driving automobiles might help spur an increase in hypercommuting and domestic cocooning. Freed from the need to pay attention while driving, people seeking to reduce their rents and mortgages might be enabled to extend their commutes, helping to create additional sprawl as they seek to leap frog their fellow citizens in pursuit of an ever more bucolic suburban home.

Any debate forum regarding these potential harms would need to have suitable precautions against the discussion being premised on the assumption that the technology is inevitable and other *scientistic* or *technocratic* framings. Ensuring that the technology's inevitability is not taken for granted would ensure a more appropriate distribution of the burden of proof between advocates and critics. Likewise, deliberation would need to be not overly tied to quantitative risk assessment or notions of "efficiency" but remain open to debates over values. As scholars who study scientific controversy have shown, a narrow focus on quantitative risk assessment or technical expertise in public deliberation often biases decision making toward the unexamined cultural values of the technoscientific experts involved.[14] Scientifically or technocratically framed debate on genetically modified crops, for instance, would primarily consider the effects on crop yields, cost, and potential human harms via ingestion. Little attention would be devoted to how such crops might reinforce a food politics based on industrial or factory farming, centralize economic power in large agricultural conglomerates and biotech firms, or otherwise influence consumers' relationship with what they eat. Hence, inquiries into autonomous automobiles would not simply examine questions concerning speed, cost effectiveness, and calculable risk of accident, but these vehicles' effect on the good life and well-being as well. Is the mode of living offered by driverless cars actually desirable or might a large constituency prefer steering toward a world less shaped by automobility?

If the decision were made to proceed with autonomously driving automobiles, governments would need to require and help fund initial precautionary measures, extensive testing, gradual scale-up, and the flexible implementation of the technology. One fairly straightforward initial precaution against sprawl would be to shift road and highway funding from gas surcharges and general taxation to a tax on vehicle miles traveled.[15] Doing so would force riders and drivers to bear more of the costs of automobility, and make it easier to implement congestion charges and other fees. By preventing the user-borne costs of automobile travel from decreasing too much, a vehicle-miles-traveled tax would help avoid certain anticommunitarian outcomes like increased hypercommuting or the abandonment of multimodal transit. Moreover, implementing such a tax on driverless cars would be simple from a technical standpoint, given that they already depend on global positioning satellites for navigation. Indeed, experiments suggest that it would be relatively straightforward to assign close to 99 percent of the tax receipts to the correct jurisdictions.[16]

Testing requirements could go far beyond the fairly narrow examinations of safety currently being conducted by Google and a few academics.[17] Evaluators would further need to consider the potential effects on individual and social well-being. Indeed, the easily recognized advantage accruing to Google, if driverless cars were implemented, is that automobile travel would become yet another corner of contemporary living colonized by digital devices and advertising-driven surveillance. Would the ability to play Candy Crush or check Facebook in a driverless car do much to counter the psychological effects of gridlock or the civic deficits produced by commuting?[18] Would companies expect greater output from their nondriving commuting workers? Would the interior environments of driverless cars encourage passengers to interact with one another, or would they exacerbate the problem of digital cocooning?

Secondary and tertiary effects of driverless cars could be discerned before too much financial and physical capital has been invested, if their broader deployment were limited to a relatively small set of cities or regions for several years or a decade. It would be far easier to change or reverse course on driverless cars if there were hundreds of thousands on roads instead of several million. Flexibility, moreover, could be maintained by not permitting hype over autonomously driven automobiles to lead to the neglect or dismantling of sidewalks, bike lanes, and public transit. Decision makers, in their excitement over driverless cars, risk repeating the same mistakes made in the mid-twentieth century when many people saw the original automobile as the end of transportation history. Might the driverless car be viewed by some observers to be a "miracle innovation" partly because it continues rather than opposes the sociotechnical momentum of the status quo of automobility? In any case, it would be wise to build in flexibility. Future generations might decide that a transportation system built around the driverless car is just as undesirable as traditional automobility. Increasing energy and resource scarcity coupled with rising global demand, barring some miraculous innovation like nuclear fusion, might render infeasible the vast communication networks, highway systems, high-end electronics manufacturing, and stable electricity grids that driverless cars need to function. Intelligent trial-and-error strategies would help ensure that driverless cars do not end up being the twenty-first-century analog to the abandoned roadside statues of Easter Island: a testament to the consequences of a myopic pursuit of unsustainable notions of progress.[19]

Besides effective deliberation, publicly accountable and fair decision-making processes, and a gradual deployment process, ITE steering could be encouraged by funding monitoring efforts and incentivizing error correction. It is already quite clear, from cases like Ford's negligence regarding the Pinto's exploding gas tank or Toyota's delayed recall on sticky accelerator pedals, that automobile technology firms will often disavow responsibility or drag their feet in dealing with the harms resulting from their products.[20] The tendency of automobile companies to shirk responsibility and fight regulation rather than engineer solutions frequently stymies sensible and largely beneficial innovations.

Governments might avoid a similar situation for driverless cars by assigning some accident liability to the firms designing the product by default, placing part of the burden of proof on producers rather than exclusively on drivers/occupants. Firms would likely respond by carrying "technical malpractice" insurance, similar to the policies doctors keep in order to compensate victims of medical mistakes. Indeed, many architects and engineers in the construction industry already carry professional liability or *errors and omissions* insurance.[21] This legal arrangement would encourage firms to more quickly correct mistakes, because they would want to keep their insurance premiums low. Some legal scholars, however, contend that potential civil penalties are weak deterrents of irresponsible design, arguing that they should be supplemented with criminal penalties for negligent engineers and corporate officers.[22]

As important as such policies would be, legal liability tends to extend only to primary undesirable consequences, namely accidents, and not to secondary and tertiary sociocultural effects. For example, it is a near certainty that autonomous automobiles would lead to job losses among taxi drivers and truckers, and in other professions. As observers of similar economic changes have noted, such "economic dislocation [generally] entails the wholesale destruction of civic networks."[23] To incentivize driverless car firms to correct the "errors" wrought on working-class families and their communities, rather than run roughshod over them, an additional sales tax on rides and driverless car purchases could be used to force consumers and producers to subsidize the welfare benefits and retraining of people put out of work. Alternatively, funds might be raised by charging driverless car firms for the access to the public wireless band upon which their innovations depend. Such charges could be scaled alongside increases in unemployment, which might incentivize some large firms to devise their own schemes to get the unemployed back to work. Finally, urban scholars and geographers would be employed to monitor changes to residency, work, and social patterns. They would be tasked to investigate questions concerning the effects of driverless cars on social inequity and the practice of community.

Companion Robots and Psychocultural Risk   Attempting to address the risks of driverless cars is similar to preventing the harms of toxic chemicals or automation. Working hard to lessen the effects on the urban environment, accidents, and job losses probably would not seem too controversial to governments and many citizens. Technologies such as companion robots, on the other hand, pose what could be called *psychocultural* risks. The possible harms have less to do with the environment or people's livelihoods than with potentially insidious changes to culture and users' psychological dispositions. A case in point is how relying on surveillance and security technologies to induce moral behavior in children, as well as to protect them from perceived outside threats, creates new risks at the same time that it fails to address the underlying social causes of violence.[24] Such techno-fixes for the problem of safety and moral development not only direct attention and money away from better social support programs, but also risk producing a false sense of security in adults, creating feelings of depersonalization and the loss of privacy in youth, and discouraging adults from intervening in children's moral development through empathic dialogue.

Psychologist Sherry Turkle has described the dangers of companion robots as emerging from the fact that they "promise a way to sidestep conflicts about intimacy" as well as the possibility that users will begin to judge their human relationships in terms of what robotic companions appear to offer.[25] Some of her interviewees expressed a preference for robotic relationships, citing how they would ostensibly have a bigger "database" from which to give "better" advice and that relationships with them would entail fewer demands and risks. Such robots, moreover, are unlikely to have subjectivities of their own; they would be programmed to be maximally compatible with the whims and desires of users.[26] In all likelihood, they would not be beings with their own imperfections, needs, and desires but objects to be consumed, hence posing particular risks for people's sexual development. Research on pornography has associated it with declines in the quality and quantity of sexual intimacy within romantic relationships and lower scores on measures of marital well-being.[27] Although couples for whom pornography use is not so damaging no doubt exist, there is good reason to worry that the promise of riskless robotic surrogates for human intimacy will not be a boon to some people's aspirations for lovingly fulfilling sexual relationships. Many partners would not want to be measured against an always ready, willing, and compliant sex robot. Lastly, as Turkle points out, the ability of robots to serve as alluring anodynes for loneliness might enable children to feel less guilty about not visiting their elderly parents, and might reduce governmental and charitable organizations' concerns about the chronically isolated.[28]

Despite these risks, there are those who want to charge forward with the development and deployment of companion and sex robots. AI expert

David Levy has written an entire book expounding the ostensible inevitability, normality, and advantages of relationships with sex robots, advantages he appeared to attribute partly to their presumed greater knowledge of the mechanics of the human orgasm.[29] MIT robotics researcher Cynthia Breazeal has contended that her work is "not about ... replacing people" even as she released a robot called JIBO that, along with other features, reads bedtime stories to owners' children.[30] To be fair, Breazeal appears more cognizant of the potential unintended consequences than advocates such as Levy. Nevertheless, her misreading of the history of automation as having merely eliminated "the jobs that people don't necessarily want to do anyway" and as "[empowering] people to do more interesting work" suggests a great deal of naiveté about the politics of technological change. If societies are to avoid the potentially undesirable unintended consequences produced by companion robots, techno-enthusiasts should not be alone at the helm—given their clear biases and conflicts of interest.

If technological societies were to intelligently steer the development of sociable robots, products such as JIBO would likely not be on the market without some form of premarket testing. Technological societies might rise to the challenge posed by such devices by instituting a version of the U.S. Food and Drug Administration (FDA) to evaluate technologies with psychocultural effects. Technologies under their purview would include devices promised to help users realize happier and less harried lives, education technologies claimed to produce smarter children, and social robots declared to "strengthen human relationships" rather than detract from them. Firms might be forced to have independent scholars perform the kind of clinical and observational studies that Sherry Turkle has done for decades, asking questions including, "How does the use of this technology influence and interface with users' expectations, dispositions, desires, and anxieties?"

Different populations undeniably embrace and use any given technology very differently. Technologies with potential psychocultural effects do not pose the same risks for everyone. Nevertheless, citizens and policy makers currently have little to no information to guide their decisions about community-influencing technologies. Making conclusions about potential psychocultural consequences, moreover, is less straightforward than determining whether an anti-inflammatory medicine contributes to heart attacks. The desirability of these technologies depends on the vision of the good life embraced by those viewing it. Robotic love advocate David Levy, for example, has seemed guided by a vision of the good sex life characterized by a more mechanistic than emotionally thick understanding of

intimacy.[31] Some of Sherry Turkle's interviewees explicitly privileged relational convenience over emotional depth in their considerations of possible
robotic companionship.[32] Nevertheless, improved premarket assessment of
these technologies, at a minimum, could enforce greater honesty by the
firms producing them. Imagine a storytelling social robot coming with a
label stating, "Studies have shown that this robot enables busy parents to
substitute time with their children with digital surrogates, which could
negatively affect the emotional depth of familial relationships. Children,
in turn, may be taught that their needs for belonging and intimacy are
best met through gadgets rather than by relating to other human beings.
They may even come to lag behind their peers in social and emotional
development."

Other aspects of an FDA-style governance might be difficult to map to
psychocultural devices. At what point would a recall of a social robot be
made? Unlike drugs such as Vioxx, which contributed to heart attacks and
strokes in users,[33] the potential harms of psychocultural devices may not
manifest in clear physiological symptoms. Moreover, due to entrenched
patterns of thought concerning "personal responsibility," people tend to
assign more blame to users of digital devices than to those making them.
Problems are taken to emerge from improper use rather than improper
design. Consider Internet sociologist Zeynep Tufekci's recent tweet: "Do all
these people who write about fake & wasteful social media ever ponder if
this relates to their choices in friends & media usage?"[34] Professor Tufekci
seemed to suggest that those who are concerned that social media too easily
encourages shallow social practices should be putting the blame on themselves, not the technology.

Clearly the relationship between design and effect is more complex for
gadgets than it is for pharmaceuticals. Users of digital devices no doubt
coshape their effects in ways unavailable to people taking medicines: a
patient's attitude about medicine is unlikely to change whether his or her
heart suffers an infarction in response to taking Vioxx. On the other hand,
people regularly struggle to follow through with their decisions regarding
the use of both substances and technologies. Consider those who struggle to abstain from alcohol or electronic gambling machines, despite the
existence of those who can more easily enjoy both with moderation.[35]
People can and do use technologies in ways that are at odds with their
longer-term or more highly valued goals and struggle to alter their
entrenched sociotechnical habits. Any argument that people's usage of
technologies "reveals" their "true preferences" or is reflective of "rational"
choosing has little logical or empirical basis. Consider a study of classroom

instant messaging, which concluded that students "seem to be aware that divided attention is detrimental to their academic achievement; however, they continue to engage in the behavior."[36] Similarly, many readers have probably experienced and regretted a night spent at home on Facebook or Netflix rather than going out to socialize. The agency of users regarding the effects of technologies on their lives is always limited and circumscribed.

The same would almost certainly be true of companion robots: Many users would enjoy them and continue to use them despite an awareness of how they conflict with other aspects of their well-being. Although technological libertarians are likely to protest, many citizens would welcome policies that provided helpful nudges toward behaviors they consider more desirable. Taxing devices that encourage cocooning and other anti- or asocial behaviors and using the funds to develop communitarian alternatives, as I have suggested throughout, would be a good first step.

Another potentially helpful move by a psychocultural technology assessment organization would be to encourage developers to make their products more flexible by allowing users to lock out some aspects of the functionality. Indeed, reflective, intelligent choice making when using cell phones and other screen technologies is frequently difficult, because they too easily cater to the users' whims and anxieties. These devices' apparent functionality hides how they persuade users toward undesired and inflexible habits. Users' own behavior reflects this fact. Consider the popular computer program *Freedom*, which allows users to shut-off their Wi-Fi access until they reboot their computers. Technology writer Evgeny Morozov has even gone so far as to lock up his Wi-Fi card in a timed safe (along with the screwdrivers he might use to circumvent the timer).[37] Users would not be driven to such lengths if their ostensibly empowering devices merely enhanced their own agency. Such features, in any case, ought to be built into risky devices. Breazeal's company, for instance, would be required to provide users with an inexpensive service through which they lock out certain functions of JIBO, such as storytelling, until the user comes back and requests that it be unlocked. Other social robots might be sold only under the condition that they be nonfunctional between the hours of five and eight, discouraging users from cocooning with them in the evening. In the same way that many European countries do not allow most stores to remain open on Sundays, recognizing that doing so encourages friends and family to gather over meals and drinks at least once a week, new gadgets could be limited so as to leave similar openings for or nudges toward embodied togetherness.

Implementing ITE Governance   The above suggestions only outline what the ITE-like governance of technologies has to offer in the abstract. Where would ITE exist? At what scale? Who would be responsible? I will not pretend that the next few paragraphs provide an entirely satisfactory answer. Nevertheless, some of the more concrete possibilities are easily imaginable by looking to the governance structures that already exist in the United States and elsewhere.

I have already suggested that something like the U.S. FDA could regulate products with psychocultural risks at a state or national scale. Yet, given the greater complexities in ascertaining psychological and cultural—as opposed to physiological—harm and the cultural barriers to regulating consumer goods outside the realm of health and human bodies, the more recently formed Consumer Financial Protection Bureau (CFPB) may be a more apt model. An institution modeled after the FDA would serve as a gatekeeper, having the power to prevent the entry of new innovations into consumer markets and take them off store shelves. Even though such capacities might be highly desirable in certain cases, it may be too politically difficult to enshrine an institution with such powers in the foreseeable future. Citizens may be too well accustomed to the status quo of innovation without permission or representation to demand it, and the resistance of powerfully connected technology firms may prove too great. An institution modeled after the CFPB, in contrast, would be tasked with educating consumers, enforcing greater transparency by firms regarding potential harms, and conducting behavioral research. Moreover, it would have the ability to use appropriate mechanisms (e.g., taxation/subsidy) to incentivize more communitarian technologies and disincentivize potentially harmful ones. Regardless, in the United States' political context, the creation of either such organization would entail an act of Congress, involving congressional approval of a presidentially appointed director. Given these political challenges and barriers, other nations might be the first place to look for the first psychocultural technology assessment institutions. Indeed, countries such as Denmark are much further along in this regard.

What about technologies whose effects on community life are not merely psychocultural but simultanously economic, material, and political? With the demise of the Office of Technology Assessment, which advised Congress on new technoscientific innovations, evaluation of technological effects happens in a far more decentralized and ad-hoc fashion.[38] Components of assessing and regulating driverless cars might fall under the purview of the Departments of Transportation, Labor, and Housing and Urban Development as well as the Federal Communications Commission, without any clear means of coordinating among these disparate agencies. The visions of individual governmental agencies are often too narrow to adequately envision the range of possible desirable and undesirable consequences. The National Highway Traffic Safety Administration is concerned primarily with the possible consequences of autonomous vehicles for individual safety, emissions, improving route planning data, and the mobility of the disabled, which are among but far from exhaust the possible sociopolitical implications of the technology.[39] Reinstituting an improved version of the Office of Technology Assessment, perhaps modeled after the Danish Board of Technology that has advised Denmark's public and parliament since 1986, would be the most obvious pathway for ensuring that some minimal level of ITE technological governance is performed.

With any sort of institutional design, the devil is in the details. Careful attention ought to be paid to establishing institutional independence. Without it the bureaucracy faces a significant risk of regulatory capture, wherein the actors being regulated by an institution manage to manipulate it to serve their own interests.[40] Consider how hydraulic fracturing methods for drilling for natural gas are not subjected to the standards of the Clean Water Act. This exception was implemented at the behest of then-vice president Dick Cheney—arguably on behalf of the oil industry, which in turn tied the hands of the Environmental Protection Agency. The FDA, moreover, is itself often criticized for being a "captured" agency and having too close of a relationship with industry.[41]

The benefit of providing a bureaucratic institution with a large amount of independence, although commonly framed as "protection from politics," is that a commitment to serve a particular set of partisan interests can be rendered obdurate. A feature that seems antidemocratic on its surface can actually enhance democracy by ensuring that the needs and wants of less empowered groups continue to be served despite opposition by political elites. As the political scientists Charles Lindblom and Edward Woodhouse have pointed out, "business groups … tend to have the advantage of better organization and finance compared with most other organized interests."[42] Hence, any organization tasked with helping to prevent consumers from becoming psychocultural or socioeconomic victims of innovation would need some assurances of financial and other kinds of independence, protecting it from capture by the powerful business firms that produce new gadgets.

ITE-style governance, at the same time, need not occur only at such large scales. Towns, cities, and regions could more often regulate the technologies existing within their borders. Doing so would have the added benefit of enhancing the practice of local political community, as do social movements more generally (See box 10.1). Such efforts might draw lessons from the Amish. Contrary to the widespread belief that they are simply technology-fearing luddites, the Amish have been practicing something similar to ITE within their district communities for generations.[43] The introduction of any technology is heavily debated and voted on by the adult residents of each district, with new technologies being given trial periods. As a result, most Amish districts far from replicate life in the seventeenth century: one sees pneumatic power tools, diesel generators for charging batteries, and neighborhood telephones alongside horse-drawn buggies and gas lighting. This selective embrace of technology reflects the cultural momentum of Amish ways of life as much as their practice of evaluating new technologies against highly revered collective values: humility, community, equality, and simplicity. Technologies, including automobiles, television, and personal phones, have been rejected because the Amish believe that they lead users to be prideful or neglectful of their ties to their families and neighbors. My point here is not that the Amish are perfect or that societies should reject electricity. Rather, I mean only that other communities could employ a similar willingness to govern technologies with regard to their values.

Rather than wait for the federal government to act, communities could begin to regulate new technologies on their own. I have already suggested that cell phone or Wi-Fi signal is one possible area: communities might designate certain areas to be free of cell phones and other devices, apart from emergency calling, to cultivate the public sociability they desire. Towns and cities might choose to prohibit driverless cars, despite whatever their state government decides. Imagine local sheriffs impounding Google's self-driving automobiles! Indeed, recent moves by cities to step in where national governments are laggard, such as Seattle's banning of plastic grocery bags and implementation of minimum wage increases, suggests a growing willingness to turn to municipal-level politics in dealing with twenty-first-century problems.[44] In any case, devolving technology assessment to the local level has the added benefit of further thickening political community. Indeed, studies of deliberative forums, namely citizen panels, find that they strengthen participants' sense of community and practice of citizenship.[45]

The barriers to acting more like the Amish, however, are significant. Localities working in such a mode risk drawing themselves into legal battles with large corporations and higher levels of government. Fortunately,

Box 10.1
From Tahrir Square to Zuccotti Park: Community, Place, and Politics

For a few months in 2011, thousands gathered daily at an obscure Manhattan park to protest American inequality. Inspired by Egypt's Arab Spring protests in Tahrir Square, the Occupy Wall Street movement went further than most demonstrations: hundreds literally occupied Zuccotti Park, overnighting in tents and sleeping bags. The occupation's level of self-organization rivaled that of many small towns. A finance working group managed a budget of eventually almost seven hundred thousand dollars. Comfort stations maintained and distributed a stock of clean clothes, blankets, and tents. Regular meals were prepared for participants and for a portion of the local homeless population. Dishes were washed and the park cleaned. A sizable library as well as a communications center offering Wi-Fi and livestreaming capabilities were located on-site. Occupiers, moreover, practiced thick political community through deliberative processes and near-consensus voting. Although eventually taken down by police, the protest has lived on through spin-off movements—including the fight for a fifteen-dollar minimum wage—and organizations such as Rolling Jubilee, which buys delinquent debts in order to forgive them.

The distinguishing feature of Occupy and Arab Spring, contrary to hype about them being Internet-based revolutions, is that they coalesced in a place. Both no doubt relied somewhat on Internet technologies and national networks of supporters. However, it was their sheer physical presence that gave them strength and worried those in power. Indeed, compare their influence and effects to the online petitions and other purely networked political activities often derided as mere "clicktivism." Occupiers hashed out strategies and aims through face-to-face deliberation, though not always constructively. Prodemocracy agitation in Egypt benefited as much from local mosques, amenable urban spaces, and the tradition of the post-Friday-prayer protest as from Twitter. Both cases illustrate how, even in the purported network society, place and copresence are still vital to social movement politics.[47]

In the eyes of some participants, the failure of the Occupy movement to sustain itself or leverage its energy into more concrete and substantive changes was partly the product of communitarian deficits. Indeed, Yotam Marom has argued that the movement "tore at the seams" with enough state pressure because participants "weren't ... grounded in communities enough." The collective struggled to productively steer internecine disputes into shared agreements and was further hampered by a "call-out" culture frequently more focused on tearing off strips than collaboratively addressing the underlying roots of perceived wrongdoings. Most of all, the protest faltered when it came to concretely outlining a shared future for both the participants and the 99 percent.[48]

> Seen from a different angle, Occupy was more than a protest: it was a prototype. If a group of people without electricity and running water can run a small tent village for months as a democratic political community in the chaos of lower Manhattan, than the same could be done for other organizations, infrastructures, and towns. Occupy's lost promise is not so much failing to sustain an exhausting long-term demonstration but that it never shifted from being a political gesture to generating the building blocks for an alternative technological society.

the Amish already provide lessons regarding how cities might respond: they have maintained a National Amish Steering Committee, which is tasked with hammering out compromises with governments concerning military service and a range of regulations that conflict with their communities' rules and ways of life.[46] For instance, mid-twentieth-century Amish objected to paying into or receiving government funded social security. In an analogous fashion, likeminded cities could band together to press against or encourage changes to federal and state rules that would otherwise stand in the way of experimentation with technological governance.

Most towns and cities, at the same time, would likely elect to be more selective than the Amish regarding the technologies they would choose to regulate. Since most localities lack the cohesiveness and insularity of most Amish settlements, many of the informal sanctions (e.g., shunning) used by the Amish to regulate personal technologies would be ineffective and probably opposed by most residents anyway. Community-level ITE would likely focus on the technologies most clearly working in the public realm, and towns and cities would discourage the use of certain personal technologies through a variety of indirect means, such as by supporting public alternatives or eliminating perverse subsidies.

Widespread disbelief in the feasibility of technological steering is likely, however, to remain a significant barrier. Officials in many localities appear more interested in the twenty-first-century equivalent of chasing smokestacks: attempting to woo creative class hipsters, rather than ensuring that technological developments do not threaten important social values. Trying to keep up with "technologically hip" cities inhibits more reflective thinking about ends and means. Citizens too frequently parrot, albeit often as a lament, the thought-stopping cliché "You can't stop progress." Equally important as the search for ways to institute ITE are interventions into the patterns of thought that can thwart even easily achievable actions.

ITE as a Set of Strategies

The benefits of an intelligent trial-and-error approach do not consist solely in the avoidance of undesirable unintended consequences but also in enhanced organizational effectiveness. ITE is not merely a mechanism for better steering the introduction of toxic chemicals, the application of genetic engineering, and the mitigation of global climate change; deficits in ITE likewise lead to cost overruns, lengthy delays, and inflexible technologies that fail to deliver on their promised benefits. Organizations that overestimate their expertise and the ability to plan for every eventuality rather than anticipate learning partly via experience are more likely to experience these results. Twentieth-century nuclear power, for instance, was not a failure merely because it resulted in disasters such as Three Mile Island, but because it was scaled up too quickly and without enough input from skeptical voices. Only after the construction of dozens of reactors was it realized that neither the plants nor the resulting electricity would be as inexpensive as hypothesized and that citizens would come to see meltdown-prone light water reactors as insufficiently safe. The Space Shuttle program not only produced the Challenger disaster but left NASA with an inflexible and expensive system of reusable shuttles and launch centers. Large initial capital investments led them to "stay the course," despite the shuttle's many inadequacies.[49] The tactics of sensible initial precautions, gradual scale-up, inexpensive trials, quick feedback, diverse participation, and shared political power are as useful for more successful innovations as they are for averting disaster. How might they guide the implementation of thick communitarian technologies?

Trial-and-Error Urbanism   Despite the promise of new urbanist designs to enhance the communality of neighborhoods, their actual implementation frequently leaves much to be desired. Indeed, critics mobilize a litany of complaints that sound eerily similar to those arising in case studies of organizational failure. First, the design process can fail to enable fair or diverse participation. Although most new urbanist developments involve a public charrette, many contend that charrettes are more often a ritualistic performance of participation than a substantial implementation of shared governance. Participation is further limited in cases where the original designer's architectural vision is written into the local code or ownership covenants.[50]

The typical new urbanist project is a massive greenfield or infill development undertaken by large institutions or firms, entailing long lead-ups, high initial capital investments, and few mechanisms for eliciting timely feedback prior to attempting to sell large swaths to potential homebuyers. Consider Mesa del Sol, a twenty-square-mile development south of Albuquerque. Taken on by Forest City Enterprises in a public-private partnership with the city and publicly subsidized through a tax-increment financing scheme,[51] the massive planned community is projected to take some forty years to complete. Major errors are already apparent. Forest City has recently tried to free itself of the project as homebuyers have proved to be apprehensive about purchasing housing in an experimental neighborhood in the middle of the New Mexico desert.[52] Likewise, the city of Edmonton appears to be forging ahead—without the involvement of a private developer—with a similarly massive twenty-five-year redevelopment of its former city airport: Blatchford. As for Mesa del Sol, one wonders if proposed sustainable and communitarian features will last over the long development period or in the face of unforeseen errors. The city has already dropped the idea of a pneumatic garbage system.[53]

Analysis of similar projects points to several foreseeable shortcomings to large new urbanist developments.[54] These planned communities tend to fail to deliver on promises of increased walking, often being "transit ready" but neither well-integrated with public transportation nor cutting back on parking enough to discourage driving. Because developers tend to be fairly risk averse when it comes to deviating from the status quo, resulting neighborhoods often fail to substantially increase densities, lending credence to the critique that new urbanism is merely more photogenic sprawl. Furthermore, new urbanist developments are usually much more expensive, reflecting higher-quality building along with the mark-up typically associated with "niche" goods. Indeed, few new urbanist developments are affordable for lower- and middle-class wage earners.[55]

Compare such results with that of a more ITE-style process implemented during the development of the pedestrian-friendly, sustainability-focused neighborhood of Quartier Vauban in Freiburg, Germany.[56] Deliberation began early on and with diverse interests represented. Citizens were not merely consulted or informed but participated continuously in the project's planning via the Forum Vauban. Experts in the planning bureau responded to citizen input, although mediation was necessary, and citizens often felt they had to "'fight' planners to convince them of the validity of their suggestions."[57] Still, the influence of traditionally powerful actors was lessened. Gradual scale-up and incremental experimentation was attained by subdividing the development into pieces that were purchased by several private developers and around forty citizen-organized "building collectives," or Baugruppen.[58] Selling parcels off to individual Baugruppen took financial pressure off the city, which otherwise would have been incentivized to offload the property as quickly as possible to a single developer.

The overall planning approach was referred to as "Learning while Planning" by locals. The chief urban planner was open to "allowing the development plan to change as a result of continuous learning and evolving standards of the Baugruppen and Forum Vauban. As goals and energy standards evolved, the city was able to incorporate these by putting new restrictions on builders via sale contracts, thus steering development."[59] Though far from perfect—power could always have been more democratically distributed and monitoring improved—the planning process at Vauban displayed a degree of preparation for learning that is incredibly rare in residential development projects of that scale.

The promise of the planning approach pursued in Vauban is apparent in how participants succeeded where similar efforts elsewhere failed. The neighborhood contains a much higher percentage of renewable energy and passive design homes as well as a much lower rate of driving and car ownership than the rest of Germany—and even compared to elsewhere in Freiburg. Indeed, citizens managed to achieve quite radical features, including designating many areas as car-free and requiring owners to pay to park their automobiles in garages at the edges of the neighborhood. Over the course of the planning process the vision for Vauban's development became less and less conventional. This starkly contrasts the North American experience with new urbanism and other alternative urban designs, where grand visions are frequently watered down as neighborhoods actually get built and sold.[60]

This result is probably not solely due to the more incremental and deliberative planning process. Certainly, Germany, and especially Freiburg, has a favorable environment for these kinds of developments. Moreover, its successes could be, at least partly, attributed to parceling out pieces of the neighborhood to soon-to-be residents to design themselves. There is an undeniable psychic value to the ability to participate in the design and construction of one's environment and dwelling space as opposed to purchasing from a list of models imagined by someone else. This no doubt can partly explain the large degree of citizen buy-in in Vauban when compared to developments such as Mesa del Sol.

A Tale of Two Groceries   Similar lessons about the utility of intelligent trial-and-error steering can be seen on a smaller scale by comparing two different New York food cooperatives. The Pioneer co-op in Troy opened to much fanfare in October 2010—after some five years of planning—in a newly renovated four thousand-square-foot building with five hundred members and forty-one workers. The Mohawk Harvest Co-op in Gloversville opened around the same time after a year of planning, crammed into an eight-hundred-square-foot rental space with only around one hundred members. Within a year the Pioneer co-op closed, having amassed some 1.9 million dollars in debt.[61] Two years after opening, the Mohawk Harvest Cooperative was consistently earning sufficient revenue to justify moving into a much larger space, continuing to operate to this day.

Given the characteristics of the two towns, one would have hardly expected Gloversville to have been more successful than Troy in producing a flourishing food cooperative. While they are both struggling postindustrial American cities, Troy boasts double the population, a college and a university, and a downtown that did not contain a full-service grocery. In contrast, there is a Price Chopper on Main Street in Gloversville, less than a mile from Mohawk Harvest, and a Hannaford not much farther away. Troy, moreover, is located in the capital district of New York (population around 1 million). Hence, the Pioneer co-op would have been able to attract shoppers from neighboring towns who might have wanted to avoid driving all the way to Albany or Schenectady to patronize a cooperative grocery.

Harvest Mohawk succeeded where the Pioneer co-op failed partly because the organizers of the former emphasized learning over time rather than attempting to imagine and implement the "perfect" cooperative grocery from the outset. This emphasis by organizers of the Harvest Mohawk Cooperative was driven largely by circumstances. While their compatriots in Troy were able to locate around 2.5 million dollars in grants, bank loans, and member investment, the organizers of Harvest Mohawk had their initial grant and loan applications rejected. They could locate only tens of thousands of dollars in member investments and a personal loan from a local retired dairy farmer.

Easy financing for the Pioneer co-op was a mixed blessing. Organizers spent five years purchasing a somewhat suitable location for around two hundred thousand dollars and then proceeded to sink hundreds of thousands of dollars into renovations and equipment prior to ever making a single sale. Indeed, a recent real estate listing for the vacant building mentioned three walk-in freezers, and the store boasted several brand new produce, dairy, and meat coolers. Along with an initially bloated payroll, an array of overhead costs including some sixty-five thousand dollars a year in interest payments alone meant that turning a profit required a level of patronage that the Pioneer co-op never saw during its first year.[62] Moreover, and despite years of planning, the organizers of the Pioneer co-op never seemed sure of their exact mission. They appeared to be trying to compete on price for conventional foodstuffs with nearby supermarkets as well as Troy's bodegas and corner beer stores, simultaneously trying to cater to customers interested in primarily organic and local products. Likely this was partly the product of the co-op's debt load, which may have encouraged managers to attempt to reach out to every possible demographic.

The development of the Pioneer co-op displayed several aspects of poor preparation for learning. High initial capital investments made the effort inflexible. The need to recoup these expenses dominated later decision making. Furthermore, important feedback concerning community needs and interest came much too late, arriving after organizers were already locked into owning an expensive building and a surplus of brand new equipment. The desire to create an ideal cooperative grocery from the outset rather than learn from experience resulted in an expensive and lengthy development process, which limited enthusiasm and led to some unforeseen and undesirable consequences. Members received numerous worried emails from managers fretting over the fact that half of the Pioneer's members were not shopping there. Lacking adequate feedback mechanisms, it was too easy for organizers to overlook the obvious reason: Troy is a college town. The five-year wait meant that many of the initial member-investors had since graduated and moved somewhere else.

In contrast, organizers of the Harvest Mohawk Cooperative moved quickly with a shoestring budget, hoping to grow revenue and get grants as they incrementally learned their community's needs and wants regarding a cooperative grocery.[63] They rented out a small, inexpensive location on Main Street, upgrading their space only after showing consistent revenue. Equipment was bought used whenever possible. Ironically, some of their more recent coolers were purchased from the failed Pioneer cooperative. Furthermore, rather than attempt to compete with Price Chopper, Harvest Mohawk organizers saw their mission primarily as a service organization for local farmers. They viewed their purpose as providing local farmers with a six-day-per-week market and, eventually, as a distributor between them and local restaurants and institutions. They arrived at this mission in part because of stakeholder surveys, from which they learned that many farmers neither desired nor considered themselves well suited to selling their own products at local farmers' markets. Given their long work hours during the growing season and the innumerable tasks still needing to be done over weekends, it is understandable that some farmers might not look forward to a Saturday morning spent bagging produce and counting change.

For all Harvest Mohawk Cooperative's ongoing successes, there have also been failures. Although organizers sought community feedback regularly, some patrons were inadvertently excluded when the grocery changed locations. Reflecting the cultural and middle-class background of volunteers and many members, the new location's interior was decorated with what could be termed a "foodie chic" aesthetic. As a result, organizers soon found that the previously small but significant population of food stamp shoppers ceased to frequent the store. Manager Chris Curro described seeing them stop at the front door, as if they were debating with themselves whether they belonged.[64] Yet if organizers had neither proceeded gradually nor monitored their efforts, they might not have ever recognized the exclusionary effects of subtle changes in store design. Hence, they were well prepared to recognize the error and begin to take corrective action.

Extending and Steering Toward ITE

Although promising, the intelligent trial-and-error framework nevertheless suffers from one major limitation and also faces a number of barriers. ITE primarily consists of a set of strategies for shaping the trajectories of emerging or developing technologies. It is a precautionary approach focused on preventing new mistakes rather than addressing old ones. This emphasis is no doubt apt, given that it is usually far less costly to fix mistakes early on. Nevertheless, as I have described throughout, several already established and entrenched technologies hinder the seven dimensions of community. Having more compact, walkable neighborhoods filled with third places and vibrant social interaction is a matter not simply of ensuring that new neighborhoods have these features but also of incrementally dismantling and reconstructing the existing transportation networks, zoning laws, and retail networks that make automobility and networked individualism nearly obligatory. How might ITE be extended to the project of eliminating or rebuilding technologies?

Despite any appearance of obduracy or fixedness, even the grandest artifice demands constant maintenance to continue functioning. Major renovations, moreover, are often incredibly expensive. After many decades, highways cannot simply be resurfaced but must be laid anew and have their large concrete supports replaced. Hence, the obduracy of many technologies, especially large infrastructures or urban spaces, is not simply a result of the expense and difficulty of alteration, for major maintenance is often little different in this regard. Rather, it is because maintenance decisions are too often routinized, leaving no opportunity to restart a process of intelligent trial and error. Major renovation and reconstruction of Interstate 787 in Albany, for example, has proceeded with too little discussion or debate regarding whether a transportation network designed in the midtwentieth century remains desirable. As soon as a few hundred million dollars are dumped into the project, short-term prospects for replacing the highway with a boulevard and opening up access to the riverfront are likely to become quite dim.

Standing in the way of substantially refashioning existing technologies are several questionable patterns of thought. No doubt, given the discomfort felt by most people when faced with uncertainty, "staying the course" has a certain psychological appeal. More onerous are arguments by market conservatives that treat past consumer and political choices as if they were sacrosanct. The current predominance of the suburban built environment, to take one example, is mistakenly believed to be the result of rational actors "voting with their feet" rather than a mixture of government subsidy, lobbying by the construction industry, racially motivated white flight, and regulatory decisions.[65] Within such a belief system, more self-conscious and deliberate steering of technology is framed as an attack on freedom writ large. Similarly, established institutions are often defended under the questionable logic that previous generations' political actors were undoubtedly more thoughtful and objective than today's activists and politicians. Consider the attitudes of some "originalist" readers of the United States' constitution, who insist on merely trying to "correctly" interpret the intentions of its writers or the then-prevalent understandings of constitutional meaning rather than debate the merits of those intentions and understandings. On a smaller scale, the ultraconservativeness of homeowners' associations regarding rule and design changes seems rooted in the belief that the original developer possessed a near-infallible understanding of how to maintain the coherency, and hence market value, of the neighborhoods.

Such ways of thinking are unlikely to disappear in the short term. Nevertheless, decision-making structures could be set up so that they are more frequently and substantively contested. Major infrastructural decisions typically have public comment periods, but these are sometimes as short as thirty days and leave too little opportunity for substantive revision. Some improvements could be achieved if state regulations forced departments of transportation and planning offices to hold public hearings for revising master planning documents whenever a major component of the infrastructure neared the end of its lifespan. Renovation would be considered by a maximum feasible diversity of stakeholders as if it were an emerging technology and, hence, as if maintaining the status quo were merely one possible trajectory among many. Similarly, federal and state law could require that zoning and planning codes as well as homeowners' association guidelines and condominium board rules contain *sunset provisions*. That is, they would be in effect for a period roughly consistent with the time in which buildings and infrastructure would not need major renovation. After that period, they would require the simple, or perhaps super, majority consent of stakeholders to be maintained. If consent is not forthcoming, then a full reappraisal of codes and rules would be required. Such laws would reflect the spirit of Thomas Jefferson's arguments for holding a constitutional convention for each new generation.[66] They would recognize that it is through not only the persistence of laws but also the obduracy of technological structures that decisions made by the dead end up ruling over the living.

What might be done to counter the tendency to make technological decisions without the conscious implementation of the trial-and-error learning necessary to reduce unintended consequences? Rarely is flexibility ensured or the pace and trajectory of innovation quickly changed in response to errors and evolving values. As I alluded to earlier, the reasons for this are clear: lack of widespread recognition of the need for intelligent trial and error. This can be partly attributed to plain old ignorance. If citizens have never been exposed to the idea or practice of politically governing technology, they can hardly be expected to advocate for it.

At the same time, a pervasive *governing mentality*—"a tacit and often ill-considered pattern of assumptions that fundamentally shapes political relationships, interactions and dialogues"[67]—concerning technology, politics, and progress appears to often stand in the way of democratically steering innovation. Specifically, many people subscribe to a *technocratic* viewpoint. They believe that social progress inevitably or even only results from the application of rationalistic means by scientific or technical experts.[68] In the mid-twentieth century, technocracy mainly manifested as the faith in bureaucratic management techniques. As is made clear by some observers' insistence that driverless cars become an area of "permissionless innovation," today's technocratic outlook centers around the belief that entrepreneurial innovators, often coming from Silicon Valley, will create "disruptive" technologies that will inevitably lead to economic growth and widely shared increases in well-being.[69]

Regardless of the form of technocratic governing mentality, the result is a culture in which it is mostly unthinkable to question the motivations, expertise, and abilities of technoscientific elites—or their affluent investors.

Democratic steering is seen, in response, as both unnecessary and undesirable, because innovation is believed to inexorably steer itself toward the best of all possible worlds. As technology scholars Jathan Sadowski and Evan Selinger have noted, "By focusing on technology as the dominant force in society—a force that progresses in inevitable ways—technocrats can justify their actions as merely being the outcome of rational, mechanical processes."[70] That is, it is imagined that technoscientific elites are not political partisans likely to steer innovations in directions where they themselves are likely to enjoy the greatest benefits but rather experts simply helping technology evolve toward a preordained, purely rational destination. Left out of this depiction are fundamental political questions: For whom is this innovation good? Who decides?[71]

Technocratic thinking remains dominant despite decades of research within STS demonstrating that it fails to correspond with reality; indeed, the previous chapters are examples of such research. Possible reasons for this include the fact that many contemporary elites benefit from this state of affairs and that media outlets, namely the Science Channel and *Wired* magazine, often put forth simplistic, almost mythical portrayals of scientific and technological advancement. Some of the blame, however, can justifiably be placed on STS scholars, who could take their work more seriously as a political and inevitably partisan activity. STS research too often resembles ecology prior to the development of conservation biology: a banal cataloguing of failures rather than proactive advocacy for a better future. There could be more counterexamples to Langdon Winner's characterization of much of the field as "blasé, depoliticized scholasticism."[72]

If the governing members of the Society for the Social Study of Science, the main organizing body for the field of STS, were to measure their success in terms of affecting public controversies involving science and technology rather than, or at least along with, conference attendance and the launching of new journals, people in the field might conduct themselves very differently. Fewer annual prizes would be awarded to largely inconsequential studies, such as a recent student prize for the analysis of the complexities of ornithological field recordings, and privilege those focusing explicitly on the barriers to a better technological civilization.[73] Furthermore, funds might be set up to support the PhD dissertations of social psychology students attempting to discern the best strategies for undermining technocratic governing mentalities, both within the classroom and in public discourse. Is it reasonable to expect intelligent trial and error (or a similar framework) to be implemented in the foreseeable future without these and similar efforts to enhance the political influence of STS research?

My goal in this chapter has been to outline how intelligent trial-and-error learning strategies could assist in the development of more communitarian technological societies. If technological societies were to embrace such strategies, emerging technologies, such as driverless cars and companion robots, would be subjected to more substantial testing and gradual scaling prior to their wider deployment. Moreover, there would be more careful monitoring and debate, in addition to stronger mechanisms for disincentivizing errors and for providing compensation to victims. In the case of companion robots, special attention would need to be paid to the tension between short-term hedonic gains and longer-term life aspirations. Companion robots could doubtlessly alleviate the pangs of loneliness for individual users, but they are probably unlikely to help people better integrate themselves into more loving and welcoming social communities of other human beings.

At the same time, ITE is not simply a strategy for governing technologies but a set of tactics aiding organizational success. Communitarian endeavors such as planned communities and food cooperatives could be more successful if they acted like the planners of Quartier Vauban or the managers of the Mohawk Harvest Cooperative: begin with diverse and public deliberation, scale up gradually, monitor efforts to correct for mistakes before they become too costly, and be prepared to change practices in response to lessons learned via experience.

Finally, intelligent trial and error could be improved by applying it to already existing technologies as much as to emerging innovations. Infrastructures such as highways and organizational technologies (e.g., zoning codes) could be made less obdurate if they were treated as if they were emerging technologies whenever some component of the sociotechnical systems they are a part of needed major renovations and repair.

Despite all its potential to help bring about more communitarian technological societies, ITE is unlikely to be applied to emerging or established technologies without addressing the wider barriers to its implementation. I have focused primarily on the cognitive barriers. Dominant patterns of thought frame technological developments as ideally free from conscious steering by governments and depict the consequences of those developments as unavoidable, albeit also sometimes lamentable. What philosopher Langdon Winner noted over thirty-five years ago continues to be largely true today: "People are often willing to make drastic changes in the way they live to accord with technological innovation at the same time they would resist similar kinds of changes justified on political grounds."[74] Given the ongoing problem of citizens being unable to realize the kinds of communities they desire—much less deal with industrial pollution, climate change, job losses from automation, and the decline or stagnation of well-being within technological societies—there may be no question more important today than how to better persuade citizens that the more responsible governance of technological innovation is both possible and desirable.

11 Toward Technology Studies That Matter

A Historical Analog

A visitor to Oneida, New York, in the mid-nineteenth century would have witnessed one of the grandest utopian social experiments of the era. The Oneida religious commune, of which only the eponymous silverware company and a few buildings remain, sought their understanding of moral perfection through a form of biblical communalism.[1] Those dwelling there were to understand themselves strictly as individuals and members of the Oneida commune. Any intermediating allegiance was strongly discouraged. To this end, children were separated from their parents and raised by others in a "children's house," and all community members were bound together through "complex marriage." Monogamous heterosexual relationships were seen as "selfish," an "egoism of two" that threatened the coherence of the whole community. The commune's leaders, having a tacit understanding of the shaping power of technologies, strove to ensure that the community's values were sociotechnically reflected and reinforced.[2] This was visible not only in their child-rearing techniques but also in their dormitory-like complexes, which housed mainly private bedrooms with a few great halls used for mealtimes and community-level socializing. Leaders were apparently so worried about the emergence of factions that new wings often contained few common spaces at all, lest they support group life that might be perceived as competing with the community as a whole.

Communality within contemporary technological societies is more like that of the Oneida Community than one might think at first glance, apart from its sexual politics, of course. There are aspects of Oneida's communal architecture that I have already advocated as desirable, such as building to encourage communal rituals and serendipitous socializing. Yet, one should not mistake the trees for the forest. If blown up to the scale of a city, nation, or world, sociality within the Oneida Community would better resemble contemporary networked individualism than thick community.

Contemporary citizens dwell in housing increasingly like the Oneida Community's private dormers. The analogous great hall, on the other hand, is more symbolic, taking the form of participation in consumerism within shopping malls, consumption of global culture through mass media, and the collective viewing of grand spectacles in the form of national or international sports and other televised events. Mass viewing and shopping is the contemporary equivalent to the mass meals of the Oneida. Most importantly, participation in or allegiance to social groups between the level of the individual and large-scale sociopolitical entities is relatively short lived. Citizens are sociotechnically encouraged to flit between transient networks of work and social life rather than root themselves in community, similar to how Oneida members were pushed to keep sexual and familial relationships brief. Citizens, moreover, are prepared to see individualism as natural through separation-oriented rearing practices—and later trained to understand themselves as members of a national-level community.

The preparation of people in technological societies to become networked individualists is as multifaceted as the Oneida Community's enculturation practices. Decades of training imparted by dominant institutionalized environments, organizations, and infrastructures have led many citizens to conceive of themselves more as atomistic nodes belonging to large-scale geopolitical entities and networks than as local community members. Market-oriented societies frame citizens as individuated consumers or shareholders, and hierarchical bureaucracies similarly arrange people as social atoms that receive services and pay fees rather than participate in governance. Undemocratic schooling steeps youth in a culture of individualized competition and prepares them to tolerate weakly democratic or even authoritarian workplaces. As a result, societies must make do with fewer skilled collaborators, negotiators of conflict, and deeply empathic human beings.

Networked individualism is similarly supported by ideas about moral perfection. According to liberal ideals of self-reliance and self-actualization, all individuals shoulder the duty to cultivate a suitable way of life for themselves.[3] Each person is wholly responsible for finding the right career and education; locating housing that fits his or her aesthetic taste, budget, and chosen lifestyle; and cultivating networks of friendship and intimate relationships. If there is a problem of social injustice, it is only that some kind of structural oppression stands in the way of the individual's moral

journey. Or, it is that the community a person has been raised in holds him
or her back. The supposed "egoism of thick community" becomes seen as
what hinders citizens' development as persons. Indeed, as political scientist David Imbroscio has noted, much of urban social policy is rooted in
the idea that social betterment will happen by helping the poor break into
more affluent neighborhoods where they can take on the culture of middle-
class networked individualism and mobility.[4] At the same time, feelings
of belonging are seen less as public goods cultivated in place but as something shrewd network operators achieve through their own private initiative. Studies of suburbia and solo dwelling uncover how the networked
moral order is frequently premised on minimizing or altogether avoiding
the relational interactions and commitments that can lead to interpersonal
conflict.[5]

Compared to the Oneida Community, networked individualism is sustained via a much broader range of technologies. Apart from the suburban
spaces and bureaucratic schools that play a similar role as the Oneida's
architecture and child-rearing practices, the social reality of networked
living is reflected and reinforced by still other infrastructures, organizations, and artifacts. Dominant economic systems nudge citizens toward
acting as atomized shoppers who seek out the best deal for themselves,
regardless of whether it takes them to Wal-Mart or Amazon rather than
to a local shop. Contemporary digital devices help support and naturalize
a form of sociality based on mixed networks of mostly unrelated proximate and distant ties, doing relatively little to assist the formation and
maintenance of communal solidarities. Companion robots and handguns, among other technologies, offer privatized, technological fixes for
collective problems. Feelings of loneliness and insecurity, in turn, can
become viewed as problems best solved by purchasing the right technology rather than by organizing to address the underlying social causes of
isolation or crime. Expectations for and ideas about community and collective action have adapted to what dominant technologies can offer as
previous practices have been forgotten or technologically legislated out of
existence.

Networked individualism has not been built into the technological context of everyday life through the decree of a charismatic leader, as was the
case in the Oneida Community. Instead innumerable technological changes
were allowed or even accelerated with too little thought regarding the possible consequences. A combination of progress myths, racism, housing shortages, a frenzied pace of highway construction, and a panoply of favorable
governmental subsidies and policies drove the bulk of twentieth-century

North American whites to "choose" the suburbs. During this process, few were thinking of the likelihood of large-scale changes to the practice of community. Similarly, infrastructures for providing electricity or Internet access were built with little to no consideration of whether they could support desirable forms of social, economic, and political community.

There remains, at the same time, little financial or regulatory support for community-scale infrastructure and a large number of barriers to economic localism, community policing, and other more communal alternatives. Regional malls and online retailers owe much of their success not simply to economies of scale but also to billions in tax breaks, the lack of policies substantively encouraging locally owned businesses and cooperatives, and the innumerable cognitive limitations and structural constraints that lead consumers to overlook the costs of automobility or the undermining of their own communities by distant, centrally controlled retail networks. Budgetary constraints, opposition by police unions, and entrenched cultural ideas about what policing "is" stand in the way of more community-oriented, less antagonistic and brutal practices by police.

Bringing It All Together

My aim has been to describe the networked individualization of community life as a case of technological lock-in.[6] Much like the sunk infrastructure, established research paradigms, and opposition by elite groups that render most nations locked into hydrocarbon-intensive economies, the cumulative entrenchment of a constellation of techniques, artifacts, organizations, and infrastructures has made the seven dimensions of thick community more difficult to practice and perhaps even to imagine. Given the predominance of suburban urban form, automobility, individualistic modes of schooling and child-rearing techniques, retail networks, and other infrastructures, technological societies would have to overcome a great deal of sociotechnical momentum to begin practicing thicker forms of community life. There is little evidence that the entrenchment of these technologies has been either consciously or democratically decided. Rather, it has more likely been the emergent result of failing to anticipate and guard against the detrimental influence of technological changes on thickly communitarian forms of life. As such, this has not merely been a case of technological lock-in but of somnambulism as well. There has been a collective failure "to recognize, debate, and address technological design as a core component in the shaping of everyday life."[7] Technological change has been treated as if it were a natural process to which citizens and societies must simply adapt.

In outlining the range of sociotechnical factors that have helped drive and sustain the thinning of community, I have sought to undermine the naturalistic portrayals of sociotechnical change too often dominating conversations about technology and community. In contrast to statements by otherwise excellent sociologists,[8] technologies supportive of networked individualism do not simply diffuse through societies but are supported and advanced through the cumulative entrenchment of compatible sociotechnical systems, composed of artifacts, infrastructures, people, policies, and institutions. Community has been thinned by technology not because technological progress demands it—not because of technological determinism—but because of ill-considered policies and patterns of thought.

I hope, moreover, that I have illustrated the utility of not getting bogged down in the question of "authentic" community. Working from a multidimensional conceptualization of community helps one recognize the existence of a wide diversity of thin and thick forms of community life. This conceptualization avoids reductionism. Admitting its multidimensional character from the outset helps prevent confusing the whole of community for its many possible parts, a confusion evident whenever some people write as if humanity were presented with a dichotomous choice between anonymous urbanity and a romantic idealization of the bucolic rural community. It is far better for citizens to think about which dimensions of communality they believe to be too thin, dedicating their efforts to realizing technologies that help to incrementally thicken those dimensions.

Furthermore, recognizing the multidimensionality of community helps move past overly simplistic critiques of it. Too often well-meaning progressives act as if thick community were immediately oppressive or pathologically exclusionary, often crudely referencing the Third Reich or other extreme examples. Rather than damn thick community, such examples are more illustrative of what can happen when the symbolic and moral aspects of communality become hypertrophied and disconnected from the practices of political community.

Attending to the seven dimensions of community, moreover, provides a much more nuanced depiction of the relationship between technological change and the practice of community. Child-rearing techniques, infrastructures, domestic technologies, schooling, and other technologies often receive far less attention and blame than digital devices and suburbia but are no less significant. Preparing citizens for the moral and political practice of community life is as important as setting up serendipitous interactions or encouraging denser social networks. The influence of domestic heating systems, similarly, can be easily overlooked if one believes community only manifests in friendship networks or cultural symbols.

Readers, I hope, will likewise recognize the utility and importance of approaching the question concerning community and technology as a political matter. Whether a particular manifestation of community is perceived as authentic or not depends on moral values that evoke profound disagreement. It is not a question that sociologists can settle for the rest of humanity, because any instantiation of community entails winners and losers. That physically proximate places for socializing have been increasingly replaced by social networking means that cyberasocials and others lose; their need for embodied interaction is not met. Children and the elderly are disproportionately harmed by neighborhoods poorly affording walking, transit, and unstructured play. Highly mobile, affluent professionals and corporate shareholders often benefit from the decline of thick, local forms of economic community, while lower- and middle-class workers and their families lose out. At the same time, the makeup of technological civilization can never be neutral with regard to the character of and possibilities for community. Partisans for thicker forms of community life must see to it that their idea of good life is technologically supported if their efforts are to be successful. Existing artifacts and sociotechnical systems will need to be partly displaced, and numerous barriers to realizing alternatives will have to be addressed.

Rather than presume away political disagreement, community scholars should focus on articulating clear distinctions between different enactments of community and deciding which ones they want their scholarship to aid. I have based my study on the observation that networked individualism is a relatively thin form of community life. Readers, obviously, are free to disagree with my priorities concerning the desirability of dense, multithreaded social ties; public forms of prosocial action; economic interdependence; and local democratic governance, but no one can think straight about the subject without recognizing that these are goods that thinner forms of community rarely deliver. I have chosen to direct my analysis toward aiding those who feel their needs for belonging and social connection are poorly served as a consequence of such prevailing limitations. In doing so I am no less a partisan than scholars whose writings naturalize, perhaps unintentionally, the status quo. Nevertheless, my analysis does differ in one very important regard: I have made my partisanship explicit.

With respect to science and technology studies (STS), this book has been unique simply by focusing on psychocultural goods, namely community. For much of the field's history, the emphasis has been on discerning the sociopolitical factors shaping the production of fact and the development of innovations rather than on the consequences of new technoscience for everyday life. Contrasting contemporary science and technology with broader notions of human well-being could constitute the better part of technology scholarship. I hope this book is a step toward that being so.

The preceding analysis, moreover, has added to the repertoire of *reconstructivist* scholarship.[9] Far too frequently STS scholars focus on outlining the social, political, and cultural factors that shape technological development, even though their methodological approaches are equally well suited for imagining how technologies could be reconstructed in more desirable ways. Similarly, many working in community studies devote most of their attention to outlining problems rather than envisioning partial solutions.[10] At some point, continually characterizing the decline of certain modes of togetherness provides marginal gains (if not negative ones) for those agitating for more communitarian societies. Such people need help imagining how the future could be different as much as they need analysis regarding what is missing in the present. Hence, I dedicated half of this book to exploring how to begin to work around sociotechnical barriers to develop and implement more thickly communitarian technologies. Every barrier marks a potential leverage point and an opportunity to envision alternatives to present trajectories. Because the technologies impinging on the practice of belonging are socially constructed, they could be reconstructed to better support the move toward thicker communities.

A broader recognition of the full range of barriers and options, moreover, makes change more imaginable rather than less. When reading over research critical of the new urbanist movement,[11] one might be tempted to interpret the cases of frequent failure as evidence that little can be feasibly done to make contemporary neighborhoods more communitarian. Yet, if one considers the full spectrum of technological and social factors that help steer residents away from neighborhood-scale community, then it is unreasonable to expect changes to urban technology to do the job on their own or very quickly. Consider the difference some twenty years made between the suburban life found by Herbert Gans, rich in community organizing and volunteering, and the suburban culture rooted in the avoidance of conflict and other people noted by M. P. Baumgartner.[12] In the same way that suburbia alone did not turn engaged citizens into more aloof networked individualists overnight, producing thick communitarians will require lengthy and multifaceted interventions that go beyond walkability and the densification of neighborhoods. Hence, advocates of thick community ought not be discouraged by the ostensible inefficacy or smallness of the effect produced by changes to a single technology or set of technologies. Rather, their thinking should be directed toward discerning which of seven factors remain missing or in need of strengthening.

Finally, I have explored what a more *intelligent trial-and-error* approach to innovations with probable consequences for the practice of community might look like.[13] The development of emerging technologies, such as companion robots or driverless cars, could be more intelligently and prudently governed with an eye toward potential risks for community life. Moreover, those in charge of managing potentially communitarian endeavors, including building a new neighborhood or starting a cooperative business, would do well to proceed gradually and to prepare to learn from mistakes. At the same time, some of the most significant barriers to thick community stem from already existing and obdurate technologies. Advocates for more responsive and democratic technological societies, therefore, ought to agitate for sunset policies and other measures that require an intelligent trial-and-error process be restarted when technologies and infrastructures begin to need major repair and renovation. The most significant obstacle to the more intelligent steering of technologies, however, is the widespread lack of recognition of the need for it. How this barrier might be lessened ought to be one of the main focuses of science and technology studies inquiry.

Apart from these scholarly contributions, my aim has been to produce analysis on the intersection of technology and community life that can be helpful to nonacademics. Of course, I have not produced anything like a how-to guide for community activists or a detailed roadmap to realizing thicker communities. Rather, I have attempted to write the kind of book on technology and togetherness that I wished had existed when I first started thinking about this issue—a book that directly implicates a wide range of techniques, artifacts, systems, and organizations and devotes considerable attention to the sociopolitical changes necessary to begin to steer technological change toward more communitarian ends. Although academic audiences may have found the large number of examples covered somewhat tedious, given that my theoretical claims could have been substantiated with far fewer, I have done so in response to my earlier frustrations as a lay reader with the paucity of examples offered by otherwise excellent books on technology and society. My hope is that the breadth of offered examples broadens the reader's understanding of the sociotechnical factors helping to sustain changes to the character of community life, and that this will help citizen-activists better recognize the multitude of potential targets for communitarian political action.

Loose Ends

There are several ways in which my inquiry into more communitarian technological societies could and ought to be advanced. To begin, the multidimensional conceptualization of community developed in chapter 3 could always be further refined, corrected, and better operationalized. I have provided more of an outline of the seven dimensions than a structural model. Those more talented than myself at developing mathematical models of social systems might be able to make sense of the myriad causal relationships and feedback loops at work in forms of social belonging. Such work would open the door for sociological empiricists to test, extend, and improve it. Finally, the logical connections between the seven dimensions could probably be more clearly articulated, especially political community's relationship to the symbolic and moral dimensions of belonging. I have mainly sought to open up conversation and problematize common assumptions. Political and social philosophers could no doubt firm up and rectify some of my theoretical conclusions.

This book has furthermore provided a broad, albeit selectively deep, account of the range of technologies impinging on the practice of community and on the barriers faced by more communitarian alternatives. Some of the important complexities of these technologies and barriers therefore could not be discussed in detail. Indeed, multiple volumes could be written just on steering toward democratic schools, community playgrounds that better support unstructured play, or mesh networks. Moreover, the preceding analysis had a decidedly North American outlook. No doubt the technologies that I have discussed have similar effects elsewhere in the world. High-rise construction and extremely long work hours no doubt dampen local social life in Asia, and the rapid embrace of digital gadgetry in the "developing world" is bound to have unanticipated consequences. Even though the general thrust of my argument has near universal applicability, there remain cross-national and cross-cultural differences to work out. My hope is that this book can add impetus for future research into specific communitarian technologies as they are used around the world.

This book has also been focused on the good of community, which is distinguishable from but not wholly independent of family structure. The technologies that affect the enactment of community simultaneously influence the character of familial life. Sociologist Ray Oldenburg noted that the growing expectation that spouses nearly completely provide for the emotional needs of their partners, what some have labeled the "greedy marriage," parallels the decline of the neighborhood-centering pub and café. In an era of relatively thin communities, other relationships have to take up the slack. Oldenburg argued that the "over-insulated togetherness" characterizing such marriages places undue pressure on spouses and contributes to relational insecurity, given that any one person is unlikely to be able to provide such all-encompassing support.[14] So far removed is community from many people's thinking today, however, that many therapists neglect to consider the cultivation of local social ties as a possible solution to the problem of a troubled marriage.

Likewise, many citizens lament how contemporary levels of job insecurity and career mobility make it difficult to live near and spend quality face-to-face time with siblings, children, grandparents, and other loved ones. At the same time, domestic arrangements that enculturate youth into networked individualism also contribute to households that often can feel more like dormitories than homes. They function more as a transfer station for family members occupied with their own private lives than as a focus for familial solidarity. Rather than passively accept the advancing networked individualization of the domestic sphere as a putative inevitability, how might technological societies provide more sociotechnical support for what could be termed "thick familial life"?

At the same time, many of the technologies thus far considered, such as walkable neighborhoods, community-scale energy, and tool libraries, can serve the goal of greater sustainability as well. Indeed, economic sociologist Juliet Schor has argued that building a sustainable economy focused on well-being rather than on production and consumption will demand turning toward more communal social structures.[15] Nevertheless, merely recognizing the potential for cross-subsidization between greater community and ecological sustainability is not enough. How might the synergies between the two be better realized? This might entail, for instance, ensuring that sustainability advocates do not lose sight of community in their zeal for less fossil fuel–dependent technologies. Renewable energy systems vary in the degree to which they facilitate thicker communities. Advocates might want to temper their excitement over the growing popularity of SolarCity's business model of large-scale corporate leasing of photovoltaic systems to individual residents. As desirable as the greater deployment of solar panels in itself might be, an overzealous embrace of such models can undermine efforts toward economically communitarian renewable energy systems. Communitarians, on the other hand, would do well to describe proposals for better places within neighborhoods for sports and unstructured play as not only community enhancing but as promoting sustainability by creating alternatives to shuttling children to a seemingly neverending progression of after-school activities.

Future research could focus on how strategies for achieving thicker communities might be synergized with efforts toward realizing other, ostensibly unrelated, outcomes. For instance, advocates for the mostly undone science of green chemistry and currently neglected technologies for chemical greening might more often work with localities—at least so long as national governments continue dragging their feet.[16] Indeed, despite the growing awareness that most citizens' homes and increasingly whole ecosystems contain known or suspected neurotoxins and carcinogens, work toward less or nontoxic alternatives has been plodding. Although gains have been made in Europe through the implementation of the REACH program, too little advancement has been made in federal or state policy in the United States regarding the harms caused by such substances or concerning moving green chemistry and chemical greening forward. The potential role of localities in better chemical governance is too frequently overlooked. Exceptions such as e-waste notwithstanding, most waste is disposed of near where the original product was consumed. Moreover, most municipalities are aching to kick-start their own homegrown high-tech industries. Yet few have considered redesigning local waste systems to separate out brown-chemical products and to charge higher fees for their disposal (as well as sufficiently high penalties for illegal dumping). Much in the same way that Evergreen Cooperatives is sustained by long-term contracts with local anchor institutions, revenues could be used to develop local cooperatives to recycle these products and/or develop alternatives. Such efforts would be enhanced to the extent that local institutions of higher education were incentivized or pressured into providing supportive expertise. Many technical universities devote substantial resources to studying far off galaxies and esoteric problems in plasma physics when they could be helping to develop technologies to fulfill local community needs.

Most of all, this book is intended to be a potential model for future research on other facets of technosocial life, especially those that are more intangible. Scholars who care about how well other valued modes of being are affected by the makeup and character of technological societies could proceed in the way I have: (1) Clarify the varying ways in which a certain mode of being is thought about and practiced; (2) distinguish different partisan positions as well as how the status quo assigns winners and losers; (3) imagine and inquire into alternatives; and (4) explore what it might take to make those alternatives more feasible. A whole range of lifeways are in need of such STS analysis. For instance, how could people in technological societies enjoy more meaningful jobs and supportive workplaces, laboring in ways that enhanced rather than worked against their overall well-being? Similarly, it is not hard to find people who express feeling overwhelmed, sleep deprived, set adrift, or otherwise dissatisfied by "modern" living. What might it take to enable a greater portion of citizens to be more fulfilled or restored by their leisure time or have less harried lives in general? How could everyday life be less colonialized by the dictates of consumer culture and its associated ills? What sociotechnical changes would permit a greater portion of the population to have the time and ability to pursue music, art, running, deep reading, gardening, and other focal practices? How could the barriers to more humanistic and less exam-oriented educational systems be lessened? What would it take to empower a greater number of people to practice more contemplative, less impaired ways of thinking?

Toward Technology Scholarship That Matters

As I argued in the previous chapter, STS scholars could more often concern themselves with the task of making a more democratic and responsible system of technological innovation easier to realize. Significant alterations to the character of everyday life are more enacted on ordinary citizens than by them, and almost four decades of STS research has not attempted hard enough to change this status quo. Indeed, most public debate concerning technoscientific innovation continues to proceed as if science and technology studies had never existed. Some might say that far too many STS scholars have squandered the privilege of a taxpayer- and tuition-funded salary or graduate stipend on projects and research questions far removed from the needs of most people. On the other hand, one might see this as an unintended consequence of STS researchers' efforts to gain recognition for the field as a legitimate and rigorous area of study, an aspiration that frequently means disavowing any sign of explicit partisanship and concerning oneself with esoteric topics.

Given the influence that science and technology have on people's well-being, one would think that the scholars studying technoscientific change would be at the forefront of inquiring into how a better world for a greater portion of humanity could be made possible. Yet, many STS articles read as if their authors were content to simply detail the political and cultural intricacies of controversies such as global climate change rather than leverage these analyses into strategies for outmaneuvering climate deniers and other opponents of desired policy changes. Consider Judy Wajcman's recent book on the interrelated technical and cultural factors leading an ever greater portion of humanity to feel "pressed for time," which in the end provides little concrete guidance concerning how things could become otherwise.[17] Though academically rigorous, engaging, and well written, her analysis could have provided more direct advice to those who hope to advance the political cause of living less harried lives. Much like STS research more generally, her underlying claim that *things could be otherwise* reads as if it were a whisper: absent analysis into how *things might actually become otherwise*, average people and policy makers are unlikely to hear it.

To be fair, a similar irrelevance infects much of the social sciences writ large. Indeed, consumer advocate and frequent presidential candidate Ralph Nader has lamented the near-total absence of academic social scientists working to provide strategies, tactics, and solutions to progressive politicians and activists.[18] Despite demands for more engaged or public forms of social science becoming increasingly commonplace, substantive changes have come painfully slowly. Hence, one of the most important avenues for future research might be the obduracy of academic practice. What are the barriers to STS producing scholarship more relevant to public problems and more strongly advancing desirable scientific and technological change? How do relatively privileged actors in the field consciously or unconsciously help to maintain the status quo? How could this social construction known as science and technology studies be otherwise?

My interest in such questions exposes my overarching motivation for writing this book. I am a partisan not only for thicker, and more democratic, forms of community but also for a more publicly relevant academy. As such, I have attempted to directly address the kinds of questions that Danish scholar Bent Flyvbjerg has described as characteristic of social sciences that matter: (1) Where do "we" seem to be headed? (2) Who gains and who loses? (3) Is it desirable? (4) What should or could be done?[19] I have not merely outlined the constellation of sociotechnical factors leading to technological societies being built for networked individualists as well as the resulting undesirable consequences, but I also have suggested policy changes and strategies for incrementally steering technological societies toward better support for thicker communities. Similar styles of studies could and should be done regarding other facets of more sensible technological societies, some of which I have already mentioned: enhanced sustainability, less inhumane policing, a more relaxed pace of living, more empathic and democratic schooling, and improved family life. I can think of no better calling for those privileged to be salaried STS scholars than to academically support the project of bringing about a more desirable technological civilization.

Notes

Chapter 1

1. Stephen Marche, "Is Facebook Making Us Lonely?" *Atlantic*, April 2, 2012, accessed January 30, 2014, http://www.theatlantic.com/magazine/archive/2012/05/is-facebook-making-us-lonely/308930.

2. Keith Hampton, "Social Media as Community," *New York Times*, June 18, 2012, accessed February 25, 2015, http://www.nytimes.com/roomfordebate/2012/02/12/theadvantages-and-disadvantages-of-living-alone/social-media-as-community.

3. Yochai Benkler, *The Wealth of Networks: How Social Production Transforms Markets and Freedom* (New Haven, CT: Yale University Press, 2006), 376.

4. Langdon Winner, *Autonomous Technology: Technics-out-of-Control as a Theme in Political Thought* (Cambridge, MA: MIT Press, 1977).

5. Miller McPherson, Lynn Smith-Lovin, and Matthew E. Bashears, "Social Isolation in America: Changes in Core Discussion Networks over Two Decades," *American Sociological Review* 71, no. 3 (2006): 353–375.

6. Keith N. Hampton, Lauren F. Sessions, and Eun Ja Her, "Core Networks, Social Isolation, and New Media," *Information, Communication, and Society* 14 no. 1 (2011): 130–155.

7. Laura A. Pratt, Debra J. Brody, and Qiuping Gu, *Antidepressant Use in Persons Aged 12 and Over: United States, 2005–2008*, NCHS Data Brief 76 (Hyattsville, MD: National Center for Health Statistics, 2011).

8. AARP, *Loneliness Among Older Adults* (Washington, DC: AARP, 2010).

9. Robert Bellah et al. *Habits of the Heart* (Berkeley: University of California Press, 1985).

10. Robert D. Putnam, *Bowling Alone: The Collapse and Revival of American Community* (New York: Simon and Schuster, 2000).

11. See April K. Clark, "Rethinking the Decline of Social Capital," *American Politics Research* 43, no. 4 (2014): 569–601; Thomas H. Sanders and Robert D. Putnam, "Still Bowling Alone? The Post-9/11 Split," *Journal of Democracy* 21, no. 1 (2010): 9–16.

12. Marc J. Dunkelman, *The Vanishing Neighbor: The Transformation of American Community* (New York: Norton, 2014).

13. Barry Wellman, "The Community Question," *American Journal of Sociology* 84, no. 5 (1979): 1201–1231; Lee Rainie and Barry Wellman, *Networked: The New Social Operating System* (Cambridge, MA: MIT Press, 2012).

14. Claude Fischer, *Still Connected: Family and Friends in America Since 1970* (New York: Russell Sage Foundation, 2011).

15. Rainie and Wellman, *Networked*, 9.

16. Eric Klinenberg, *Going Solo: The Extraordinary Rise and Surprising Appeal of Living Alone* (New York: Penguin Press, 2012), 18.

17. Sara Konrath, "The Empathy Paradox: Increasing Disconnection in the Age of Increasing Connection," in *Handbook of Research on Technoself: Identity in a Technological Society*, ed. R. Luppicini (Hershey, PA: IGI Global, 2013), 204–228.

18. Carolyn Marvin, *When Old Technologies Were New* (New York: Oxford University Press, 1988), 235.

19. See Thomas P. Hughes, "Technological Momentum," in *Does Technology Drive History?*, ed. Merritt Roe Smith and Leo Marx (Cambridge, MA: MIT Press, 1994), 33–41; Richard Perkins, "Technological Lock-In," *The Internet Encyclopedia of Ecological Economics* (International Society for Ecological Economics, 2003), accessed December 8, 2014, http://isecoeco.org/pdf/techlkin.pdf.

20. See Winner, *Autonomous Technology*.

21. John G. Bruhn and Stewart Wolff, *The Roseto Story: An Anatomy of Health* (Norman: University of Oklahoma Press, 1979); Putnam, *Bowling Alone*; Susan Pinker, *The Village Effect* (New York: Spiegel and Grau, 2014).

22. Juliet Schor, *Plenitude: The New Economics of True Wealth* (New York: Penguin, 2010); Ozzie Zehner, *Green Illusions* (Lincoln: University of Nebraska Press, 2012); Gabor Zovanyi, *No-Growth Imperative: Creating Sustainable Communities Under Ecological Limits to Growth* (New York: Routledge, 2013).

23. Hannah Arendt, *The Human Condition* (Chicago: University of Chicago Press, 1958); Benjamin Barber, *Strong Democracy* (Berkeley: University of California Press, 1984).

24. Robert Nisbet, *The Quest for Community* (New York: Oxford University Press, 1953; Wilmington, DE: ISI Books, 2010).

25. See Andrew Szasz, *Shopping Our Way to Safety* (Minneapolis: University of Minnesota Press, 2007); Evgeny Morozov, *The Net Delusion* (New York: Public Affairs, 2011).

26. Rainie and Wellman, *Networked*, 272.

27. *The Age of Loneliness*, directed by Sue Bourne (London, UK: Wellpark Productions, 2016), online video, http://www.bbc.co.uk/programmes/b06vkhr5.

28. Anique Hommels, *Unbuilding Cities: Obduracy in Urban Socio-Technical Change* (Cambridge, MA: MIT Press, 2005).

29. Thomas Gieryn, "What Buildings Do," *Theory and Society* 31 (2002): 35.

30. See Edward J. Woodhouse, "(Re)Constructing Technological Society by Taking Social Construction Even More Seriously," *Social Epistemology* 19, no. 2–3 (2005): 199–223.

31. Winner, *Autonomous Technology*, 325.

32. Sherry Turkle, *Alone Together* (New York: Basic Books, 2011).

33. Edward J. Woodhouse and David Collingridge, "Incrementalism, Intelligent Trial-and-Error, and the Future of Political Decision Theory," in *An Heretical Heir of the Enlightenment*, ed. Harry Redner (Boulder, CO: Westview Press, 1993), 131–154.

34. See Wiebe E. Bijker, Thomas P. Hughes, and Trevor J. Pinch, eds., *The Social Construction of Technological Systems* (Cambridge, MA: MIT Press, 1987); Wiebe E. Bijker and John Law, eds., *Shaping Technology/Building Society: Studies in Sociotechnical Change* (Cambridge, MA: MIT Press, 1992).

35. Putnam, *Bowling Alone*, 17.

Chapter 2

1. Donald MacKenzie, *Inventing Accuracy: A Historical Sociology of Nuclear Missile Guidance* (Cambridge, MA: MIT Press, 1993); Bruno Latour, *Aramis, or the Love of Technology* (Cambridge, MA: Harvard University Press, 1993); Harry Collins, *Gravity's Shadow: The Search for Gravitational Waves* (Chicago: University of Chicago Press, 2004).

2. See Peter L. Berger and Thomas Luckmann, *The Social Construction of Reality* (New York: Doubleday Anchor Books, 1966).

3. Leonard Nevarez, *Pursuing Quality of Life* (New York: Routledge, 2011), 225.

4. Robert D. Putnam, *Bowling Alone: The Collapse and Revival of American Community* (New York: Simon and Schuster, 2000).

5. Harold Lasswell, *Politics: Who Gets What, When, and How?* (New York: McGraw-Hill, 1936; New York: Meridian, 1958).

6. Robert Lane, *The Loss of Happiness in Market Democracies* (New Haven, CT: Yale University Press, 2000), 108.

7. Barry Wellman, *The Community Question Re-evaluated.* (Toronto: Centre for Urban and Community Studies, 1987); Wellman, "From Little Boxes to Loosely-Bounded Networks: The Privatization and Domestication of Community," in *Sociology for the Twenty-First Century*, ed. Janet Abu-Lughod (Chicago: University of Chicago Press, 1999), 94–114.

8. Michael V. Arnold, "The Concept of Community and the Character of Networks," *Journal of Community Informatics* 3, no. 2 (2007), accessed April 1, 2015, http://ci-journal.net/index.php/ciej/article/view/327/315.

9. John G. Bruhn and Stewart Wolff, *The Roseto Story: An Anatomy of Health* (Norman: University of Oklahoma Press, 1979).

10. Susan Pinker, *The Village Effect* (New York: Spiegel and Grau, 2014).

11. See Bernice A. Pescosolido and Beth A. Rubin, "The Web of Group Affiliations Revisited: Social Life, Postmodernism, and Sociology," *American Sociological Review* 65, no. 1 (2000): 52–67; Ray Oldenburg, *The Great Good Place.* (1989; repr., Cambridge, MA: Da Capo Press, 1999), 264–269.

12. Quoted in Robert Sullivan, "The Real Park Slope Co-op," *New York Magazine*, November 1, 2009, accessed February 25, 2015, http://nymag.com/realestate/features/61743.

13. Kirsten LaValley, "Are You Lonely, Mama?" *Huffington Post*, June 13, 2014, accessed February 25, 2015, http://www.huffingtonpost.com/kristenlavalley/are-you-lonely-mama_b_5489301.html.

14. Zeynep Tufekci and Matthew E. Brashears, "Are We All Equally at Home Online?" *Information, Communication, and Society* 17, no. 4 (2014): 486–502.

15. See John Donvan and Caren Zucker, *In a Different Key: The Story of Autism* (New York: Crown, 2016), 548.

16. John Donvan and Caren Zucker, "Autism's First Child," *Atlantic*, October, 2010, accessed June 7, 2016. http://www.theatlantic.com/magazine/archive/2010/10/autisms-first-child/308227.

17. Langdon Winner, *Autonomous Technology: Technics-out-of-Control as a Theme in Political Thought* (Cambridge, MA: MIT Press, 1977); Langdon Winner, "Do Artifacts Have Politics?" *Daedalus* 109, no. 1 (1980): 121–136.

18. See Oldenburg, *Great Good Place*, 295–296.

19. Langdon Winner, "Technology as Forms of Life," in *Readings in the Philosophy of Technology,* ed. D. M. Kaplan (Lanham, MD: Rowman and Littlefield, 2004), 104.

20. Oldenburg, *Great Good Place,* 265.

21. Barry Wellman et al., "The Social Affordances of the Internet for Networked Individualism," *Journal of Computer Mediated Communication* 8, no. 3 (2003), accessed December 6, 2016, http://onlinelibrary.wiley.com/doi/10.1111/j.1083-6101.2003 .tb00216.x/full.

22. Lee Rainie and Barry Wellman, *Networked: The New Social Operating System* (Cambridge, MA: MIT Press, 2012), 297.

23. Winner, "Do Artifacts Have Politics?"

24. See Bradford C. Snell, *American Ground Transport,* Report to the Committee of the Judiciary (Washington, DC: United States Government Printing Office, 1974); Martha J. Bianco, "The Decline of Transit: A Corporate Conspiracy or Failure of Public Policy? The Case of Portland, Oregon," *Journal of Policy History* 9, no. 4 (1997): 450–474.

25. Rainie and Wellman, *Networked,* 255.

26. Eric Klinenberg, *Going Solo: The Extraordinary Rise and Surprising Appeal of Living Alone* (New York: Penguin, 2012), 221.

27. See Richard E. Sclove, *Democracy and Technology* (New York: Guilford Books, 1995).

28. Winner, *Autonomous Technology.*

29. Putnam, *Bowling Alone.*

30. Winner, *Autonomous Technology,* 229.

31. Herbert J. Gans, *The Levittowners: Ways of Life and Politics in a New Suburban Community* (New York: Pantheon Books, 1967).

32. Oldenburg, *Great Good Place,* 265.

33. Michele A. Willson, "Being-Together: Thinking through Technologically Mediated Sociality and Community," *Communication and Critical/Cultural Studies* 9, no. 3 (2012): 279–297.

34. Klinenberg, *Going Solo,* 93.

35. Ibid., 83.

36. Robert Bellah et al. *Habits of the Heart* (Berkeley: University of California Press, 1985).

37. Klinenberg, *Going Solo,* 231.

38. Keith Hampton, "Social Media as Community," *New York Times*, June 18, 2012, accessed February 25, 2015, http://www.nytimes.com/roomfordebate/2012/02/12/theadvantages-and-disadvantages-of-living-alone/social-media-as-community.

39. Josef Keulartz et al., "Ethics in Technological Culture," *Science, Technology, and Human Values* 29, no. 1 (2004): 11.

40. Wellman, *Community Question Re-evaluated*.

41. Barry Wellman, *The Persistence and Transformation of Community*. Report to the Law Commission of Canada, October 30, 2001, 8, accessed April 1, 2015, http://groups.chass.utoronto.ca/netlab/wp-content/uploads/2012/05/The-Persistence-and-Transformation-of-Community-From-Neighbourhood-Groups-to-Social-Networks.pdf.

42. Rainie and Wellman, *Networked*, 255.

43. Zeynep Tufekci, "The Social Internet: Frustrating, Enriching, but Not Lonely," *Public Culture* 26, no. 1 (2014): 21.

44. Steven Brint, "Gemeinschaft Revisited: A Critique and Reconstruction of the Community Concept," *Sociological Theory* 19, no. 1 (2001): 8.

45. Ronald Beiner, *What's the Matter with Liberalism?* (Berkeley: University of California Press, 1992).

46. Michael. J. Sandel, "The Procedural Republic and the Unencumbered Self," *Political Theory* 12, no. 1 (1984): 81–96.

47. Rainie and Wellman, *Networked*, 125; Barry Wellman, "The Community Question," *American Journal of Sociology* 84, no. 5 (1979); Klinenberg, *Going Solo*, 104. An assumed philosophically liberal framework visibly permeates much of the contemporary social science on community.

48. Thad Williamson, David Imbroscio, and Gar Alperovitz, *Making a Place for Community* (New York: Routledge, 2002), 6.

49. Charles E. Lindblom, "Who Needs What Social Research for Policymaking?" *Science Communitcation* 7, no. 4 (1986): 345–366.

Chapter 3

1. See Roy F. Baumeister and Mark R. Leary, "The Need to Belong: Desire for Interpersonal Attachments as a Fundamental Human Motivation," *Psychological Bulletin* 17, no. 3 (1995): 497–529; John T. Cacioppo and William Patrick, *Loneliness: Human Nature and the Need for Social Connection* (New York: Norton, 2008).

2. Craig J. Calhoun, "Community: Toward a Variable Conceptualization for Comparative Research," *Social History* 5, no. 1 (1980): 1051–1029; Robert D. Putnam,

Bowling Alone: The Collapse and Revival of American Community (New York: Simon and Schuster, 2000).

3. Larry Lyon, *The Community in Urban Society* (Prospect Heights, IL: Waveland Press, 1999); Georg Simmel, "The Metropolis and Mental Life," in *The Blackwell City Reader*, 2nd ed., ed. Gary Bridge and Sophie Watson (Malden, MA: Blackwell, 2010), 103–110.

4. Herbert J. Gans, *The Urban Villagers* (New York: Free Press, 1962); Mary Malone, "The Health Experience of Irish People in a North West London 'Community Saved,'" *Community, Work, and Family* 4, no. 2 (2001): 177–196.

5. Herbert J. Gans, "Urbanism and Surbanism as Ways of Life," in *North American Suburbs: Politics, Diversity, and Change,* ed. John Kramer (Berkeley, CA: Glandessary Press, 1972), 31–51.

6. Barry Wellman, "The Community Question," *American Journal of Sociology* 84, no. 5 (1979), 1201–1231; Barry Wellman, "From Little Boxes to Loosely-Bounded Networks: The Privatization and Domestication of Community," in *Sociology for the Twenty-First Century,* ed. Janet Abu-Lughod (Chicago: University of Chicago Press, 1999), 94–114.

7. Margaret S. Clark and Judson R. Mills, "A Theory of Communal (and Exchange) Relationships," in *Handbook of Theories of Social Psychology*, vol. 2, ed. Paul A.M. Van Lange et al. (Thousand Oaks, CA: Sage, 2012), 232–250.

8. Putnam, *Bowling Alone*, 19–21.

9. Daniel Kemmis, *Community and the Politics of Place* (Norman: University of Oklahoma Press, 1990).

10. Freuchen as quoted in David Graeber, "On the Moral Grounds of Economic Relations," *Journal of Classical Sociology* 14, no. 1 (2014): 75.

11. Allen M. Omoto and Mark Snyder, "Influences of Psychological Sense of Community on Voluntary Helping and Prosocial Action," in *The Psychology of Prosocial Behavior*, ed. Stefan Stürmer and Mark Snyder (Malden, MA: Blackwell, 2010), 223–243.

12. Jesse Tatum, *Muted Voices: The Recovery of Democracy in the Shaping of Technology* (Cranbury, NJ: Associated University Presses, 2000).

13. David Graeber, *Debt: The First 5,000 Years* (Brooklyn, NY: Melville House, 2011), 105.

14. Robert Bellah et al. *Habits of the Heart.* (Berkeley: University of California Press, 1985), 128–130.

15. Quoted in Joshua Bright, "America: Single, and Loving It," *New York Times*, February 10, 2012, accessed April 13, 2014, http://www.nytimes.com/2012/02/12/fashion/America-Single-and-Loving-It.html.

16. Rochelle R. Côté, Gabriele Plickert, and Barry Wellman, "Does *The Golden Rule* Rule?" in *Contexts of Social Capital*, ed. Ray-May Hsung, Nan Lin, and Ronald Breiger (New York: Routledge, 2009), 49.

17. Gina Tron, "I Got Raped, Then My Problems Started," *Vice Magazine*, June 27, 2013, accessed August 1, 2013, http://www.vice.com/read/i-got-raped-then-my-problems-started.

18. John L. Locke, *The De-Voicing of Society* (New York: Simon and Schuster, 1998).

19. Janet A. Flammang, *The Taste for Civilization: Food, Politics, and Civil Society* (Urbana: University of Illinois Press, 2009); also see Albert Borgmann, *Real American Ethics* (Chicago: University of Chicago Press, 2006).

20. Ray Oldenburg, *The Great Good Place* (1989; repr., Cambridge, MA: Da Capo Press, 1999).

21. See Richard E. Sclove, *Democracy and Technology* (New York: Guilford Books, 1995).

22. Flammang, *The Taste for Civilization*, 11–12.

23. Howard Rheingold, "What the WELL's Rise and Fall Tell Us about Online Community," *Atlantic*, July 6, 2012, accessed April 13, 2014, http://www.theatlantic.com/technology/archive/2012/07/what-the-wells-rise-and-fall-tell-us-about-online-community/259504.

24. Zeynep Tufekci and Matthew E. Brashears, "Are We All Equally at Home Online?" *Information, Communication, and Society* 17, no. 4 (2014): 486–502.

25. Dan Buettner, "The Island Where People Forget to Die," *New York Times Magazine*, October 24, 2012, accessed June 13, 2016. http://www.nytimes.com/2012/10/28/magazine/the-island-where-people-forget-to-die.html; Andrew Anthony, "The Island of Long Life," *Guardian*, May 31, 2013, accessed June 13, 2016, https://www.theguardian.com/world/2013/may/31/ikaria-greece-longevity-secrets-age.

26. Benedict R. Anderson, *Imagined Communities* (New York: Verso, 1991), 6.

27. Ibid.

28. Craig J. Calhoun, "Imagined Communities and Indirect Relationships: Large-Scale Social Integration and the Transformation of Everyday Life," in *Social Theory for a Changing Society*, ed. Pierre Bourdieu and James S. Coleman (Boulder, CO: Westview Press, 1991), 95–120.

29. Joseph Edward Campbell, "Virtual Citizens or Dream Consumers," in *Queer Online*, ed. Kate O'Riordan and David J. Philips (New York: Peter Lang, 2007), 177–196.

30. David Proctor, *Civic Communion: The Rhetoric of Community Building* (Lanham, MD: Rowman and Littlefield, 2005).

31. See Bellah et al., *Habits of the Heart*; Nina Eliasoph, *Avoiding Politics: How Americans Produce Apathy in Everyday Life* (New York: Cambridge University Press, 1998).

32. Anthony P. Cohen, *The Symbolic Construction of Community* (New York: Tavistock Publications, 1985), 16.

33. Steven Miles, *Spaces for Consumption* (Thousand Oaks, CA: Sage, 2010).

34. James R. Beniger, "Personalization of Mass Media and the Growth of Pseudo-community," *Communication Research* 14, no. 3 (1987): 352–371.

35. See Miles, *Spaces for Consumption*.

36. See James Howard Kunstler, *The Geography of Nowhere* (New York: Simon and Schuster, 1993).

37. Marc Augé, *Non-Places: An Introduction to Supermodernity* (1995; repr., Brooklyn, NY: Verso, 2008), 81.

38. John F. Freie, *Counterfeit Community: The Exploitation of Our Longings for Connectedness* (Lanham, MD: Rowman and Littlefield, 1998).

39. Seymour B. Sarason, *The Psychological Sense of Community: Prospects for a Community Psychology* (San Francisco, CA: Jossey-Bass, 1976), 1.

40. David W. McMillan and David M. Chavis, "Sense of Community," *Journal of Community Psychology* 14, no. 1 (1986); Omoto and Snyder, "Influences of Psychological Sense of Community."

41. Thad Williamson, *Sprawl, Justice, and Citizenship: The Civic Costs of the American Way of Life* (New York: Oxford University Press, 2010), 159–161.

42. Putnam, *Bowling Alone*.

43. Rheingold, "WELL's Rise and Fall."

44. See Jared Diamond, *The World until Yesterday* (New York: Viking, 2012).

45. Richard Wilk, "Culture and Energy Consumption," in *Energy: Science, Policy and the Pursuit of Sustainability*, ed. Robert Bent, Randall Baker, and Lloyd Orr (Washington, DC: Island Press, 2002), 109–130.

46. Stephen A. Marglin, *The Dismal Science: How Thinking Like an Economist Undermines Community* (Cambridge, MA: Harvard University Press, 2008), 9.

47. Robert H. Frank, Thomas Gilovich, and Dennis T. Regan, "Does Studying Economics Inhibit Cooperation?" *Journal of Economic Perspectives* 7, no. 2 (1993): 1381–1390; Kathleen D. Vohs, Nicole L. Mead, and Miranda R. Goode, "The Psychological Consequences of Money," *Science* 314, no. 5802 (2006); Uri Gneezy and Aldo Rustichini, "A Fine is a Price," *Journal of Legal Studies* 29 (2000).

48. Quoted in *Today*, "Free Amazon Price Check App (Instantly Compare In-Store and Online Amazon Prices)," *Hip2Save*, December 7, 2011, accessed August 1, 2013, http://hip2save.com/2011/12/07/free-amazon-price-check-app-instantly-compare-in-store-online-amazon-prices.

49. Charles Duhigg and Keith Bradsher, "How the U.S. Lost Out on iPhone Work," *New York Times*, January 21, 2012, accessed April 13, 2014, http://www.nytimes.com/2012/01/22/business/apple-america-and-a-squeezed-middle-class.html.

50. Jerome F. Scott and R. P. Lynton, *The Community Factor in Modern Technology* (Paris: United Nations Educational, Scientific, and Cultural Organization, 1952).

51. Thad Williamson, David Imbroscio, and Gar Alperovitz, *Making a Place for Community* (New York: Routledge, 2002), 312.

52. Shigehiro Oishi et al., "The Socioeconomic Model of Procommunity Action," *Journal of Personality and Social Psychology* 93, no. 5 (2007): 831–844; Shigehiro Oishi and Jason Kisling, "The Mutual Constitution of Residential Mobility and Individualism," in *Understanding Culture*, ed. Robert S. Wyer, Chi-yue Chiu, and Ying-yi Hong (New York: Psychology Press, 2009), 223–238.

53. Gerard Delanty, *Community* (New York: Routledge, 2003), 71.

54. Russell Belk, "Sharing Versus Pseudo-Sharing in Web 2.0," *Anthropologist* 18, no. 1 (2014): 7–23.

55. See Benjamin Barber, *Strong Democracy* (Berkeley: University of California Press, 1984); Philip Selznick, *The Communitarian Persuasion* (Baltimore, MD: Johns Hopkins University Press, 2002).

56. Robert Nisbet, preface to the 1970 edition of *The Quest for Community* (New York: Oxford University Press, 1953; Wilmington, DE: ISI Books, 2010), xxx.

57. Alasdair MacIntyre, *After Virtue*, 2nd ed. (Notre Dame, IN: University of Notre Dame Press, 1984), 250.

58. Elinor Ostrom, *Governing the Commons: The Evolution of Institutions for Collective Action* (Cambridge, MA: Cambridge University Press, 1990).

59. Ibid., 143–146.

60. See Wendell Berry, *What Matters? Economics for a Renewed Commonwealth* (Berkeley, CA: Counterpoint, 2010); Daniel Kemmis, *Community and the Politics of Place* (Norman: University of Oklahoma Press, 1990); Paul B. Thompson, *The*

Agrarian Vision: Sustainability and Environmental Ethics (Lexington: University of Kentucky Press, 2010).

61. Diamond, *World until Yesterday*.

62. Barber, *Strong Democracy*, 178.

63. Sclove, *Democracy and Technology*.

64. Eric Klinenberg, *Going Solo: The Extraordinary Rise and Surprising Appeal of Living Alone* (New York: Penguin, 2012).

65. Quoted in Paul Kennedy (host), "Never in Anger, Part 1," *CBC Ideas*, Audio podcast, October 11, 2011, accessed April 1, 2015, http://www.cbc.ca/player/play/2263114454.

66. Karen Nakamura, *A Disability of the Soul* (Ithaca, NY: Cornell University Press, 2013); *Bethel: Community and Schizophrenia in Northern Japan*, directed by Karen Nakamura (Charleston, SC: CreateSpace, 2009).

67. Michael J. Sandel, *Liberalism and the Limits of Justice* (1982; repr., New York: Cambridge University Press, 1998), 150.

68. Bruce Bimber, "The Internet and Political Transformation: Populism, Community, and Accelerated Pluralism," *Polity* 31, no. 1 (1998): 148.

69. Stephen Vaisey, "Structure, Culture, and Community: The Search for Belonging in 50 Urban Communes," *American Sociological Review* 72, no. 6 (2007): 851–873.

70. Barber, *Strong Democracy*.

71. Nisbet, *Quest for Community*.

72. Ronald Beiner, *What's the Matter with Liberalism?* (Berkeley: University of California Press, 1992).

73. Jayme Poisson, "Parents Keep Child's Gender Secret," *Toronto Star*, May 21, 2011, accessed April 13, 2014, https://www.thestar.com/life/parent/2011/05/21/parents_keep_childs_gender_secret.html.

74. See Hazel R. Markus and Barry Schwartz, "Does Choice Mean Freedom and Well-Being?" *Journal of Consumer Research* 37, no. 2 (2010): 344–355; Hazel R. Markus and Shinobu Kitayama, "Cultures and Selves: A Cycle of Mutual Constitution," *Perspectives on Psychological Science* 5, no. 4 (2010): 420–430.

75. Gabriela Coleman, *Coding Freedom* (Princeton, NJ: Princeton University Press, 2013), 14.

76. Charles Taylor, *Modern Social Imaginaries* (Durham, NC: Duke University Press, 2004).

77. *Pleasantville*, directed by Gary Ross (Los Angeles, CA: New Line Cinema, 1998).

78. Vaisey, "Structure, Culture, and Community."

79. John G. Bruhn and Stewart Wolff, *The Roseto Story: An Anatomy of Health* (Norman: University of Oklahoma Press, 1979); Graeber, "Moral Grounds of Economic Relations," 75.

80. Steven Brint, "Gemeinschaft Revisited: A Critique and Reconstruction of the Community Concept," *Sociological Theory* 19, no. 1 (2001): 1–23.

81. Thomas Wimark, "Beyond Bright City Lights: The Migration Patterns of Gay Men and Lesbians" (PhD diss., Stockholm University, 2014); Hans W. Kristiansen, interview by Carlos Motta, *We Who Feel Differently*, November 11, 2009, accessed February 12, 2015, http://wewhofeeldifferently.info/interview.php?interview=62; Siw Ellen Jakobsen, "Gay Swedes Don't Flee Rural Communities," *ScienceNordic*, July 21, 2014, accessed February 12, 2015, http://sciencenordic.com/gay-swedes-don%E2%80%99t-flee-rural-communities; Nadine Hubbs, *Rednecks, Queers, and Country Music* (Berkeley: University of California Press, 2014).

82. Richard Sennett, "The Myth of the Purified Community," in *The Community Development Reader,* ed. James DeFilippis and Susan Saegert (New York: Routledge, 2008), 174–180.

83. Walter L. Williams, "Persistence and Change in the Berdache Tradition Among the Contemporary Lakota Indians," *Journal of Homosexuality* 11, no. 3–4 (1986): 191–200.

84. Chris Lehman, "Real Life Story of Small Town Mayor to Hit Seattle Stage," *Northwest News Network*, June 11, 2013, accessed October 16, 2014, http://www.opb.org/news/article/npr-real-life-story-of-small-town-mayor-to-hit-seattle-stage.

85. Quoted in Walter Hooper, ed., *The Collected Letters of C. S. Lewis*, vol. 3 (New York: Harper-Collins, 2007), 68.

86. See Iris Marion Young, "The Ideal of Community and the Politics of Difference," *Social Theory and Practice* 12, no. 1 (1986): 1–26. A major oversight in Young's analysis is what would be done with all those who do not share her vision of the good life as a form of semianonymous and highly mobile urban individualism.

87. Barber, *Strong Democracy*.

88. Young, "The Ideal of Community."

89. Judith Garber, "Defining Feminist Community: Place, Choice, and the Urban Politics of Difference," in *The Community Development Reader,* 2nd ed., ed. J. DeFilippis and S. Saegert (New York: Routledge, 2013), 343.

90. James DeFilippis and Peter North, "The Emancipatory Community? Place, Politics, and Collective Action in Cities," in *The Emancipatory City?: Paradoxes and Possibilities,* ed. L. Lees (London: Sage, 2004), 72–89.

91. Hannah Arendt, *The Human Condition* (Chicago: University of Chicago Press, 1958).

92. David Brain, "From Good Neighborhoods to Sustainable Cities: Social Science and the Social Agenda of New Urbanism," *International Regional Science Review* 28, no. 2 (2005): 217–238.

93. See Charles E. Lindblom and Edward J. Woodhouse, *The Policy-Making Process*, 3rd ed. (Englewood Cliffs, NJ: Prentice Hall).

Chapter 4

1. Important distinctions should be made between the different approaches of philosophers of technology, from Langdon Winner and Albert Borgmann to Peter-Paul Verbeek, but I do not discuss them here. In this and the following chapters I simply begin with the assumption that technologies are nonneutral with regard to the social consequences of their eventual use. Of course they do not wholly determine social life, but there is an often recognizable bias to how technologies mediate reality for their users. Such a bias can be opposed and sometimes overcome, but usually in situations where users have other (e.g., political, economic, cognitive) resources to support their intentions.

2. See Peter-Paul Verbeek, *Moralizing Technology: Understanding and Designing the Morality of Things* (Chicago: University of Chicago, 2011).

3. Langdon Winner, "Technology as Forms of Life," in *Readings in the Philosophy of Technology*, ed. D. M. Kaplan (Lanham, MD: Rowman and Littlefield, 2004), 103–113.

4. Albert Borgmann, *Technology and the Character of Contemporary Life* (Chicago: University of Chicago Press, 1984).

5. Trevor J. Pinch and Wiebe E. Bijker, "The Social Construction of Facts and Artifacts," in *The Social Construction of Technological Systems*, ed. Wiebe E. Bijker and Trevor Pinch (Cambridge, MA: MIT Press, 2012), 11–44.

6. Albert Borgmann, *Real American Ethics* (Chicago: University of Chicago Press, 2006).

7. See Richard E. Sclove, *Democracy and Technology* (New York: Guilford Books, 1995), 62.

8. Howard Gillette, *Civitas by Design* (Philadelphia: University of Pennsylvania Press, 2010).

9. Ray Oldenburg, *The Great Good Place* (1989; repr., Cambridge, MA: Da Capo Press, 1999), 265.

10. Kevin Lynch, *Good City Form* (Cambridge, MA: MIT Press, 1984).

11. M. P. Baumgartner, *The Moral Order of a Suburb* (New York: Oxford University Press, 1988), 9.

12. Herbert J. Gans, *The Levittowners: Ways of Life and Politics in a New Suburban Community* (New York: Pantheon Books, 1967).

13. See Jeff Wiltse, *Contested Waters* (Chapel Hill: University of North Carolina Press, 2007); Richard Louv, *The Last Child in the Woods* (Chapel Hill, NC: Algonquin Books, 2008), 127–129.

14. Kenneth A. Scherzer, *The Unbounded Community* (Durham, NC: Duke University Press, 1992).

15. Christopher T. McCahill and Norman W. Garrick, "Influence of Parking Policy on Built Environment and Travel Behavior in Two New England Cities, 1960 to 2007," *Transportation Research Record* 2187 (2010): 123–130.

16. Pu Miao, "Deserted Streets in a Jammed Town: The Gated Community in Chinese Cities and Its Solution," *Journal of Urban Design* 8, no. 1 (2003): 45–66.

17. Jane Jacobs, *The Death and Life of Great American Cities* (New York: Random House, 1961).

18. Herbert J. Gans, *The Urban Villagers* (New York: Free Press, 1962), 15.

19. Janice Perlman, *Favela: Four Decades of Living on the Edge of Rio de Janeiro* (New York: Oxford University Press, 2010); Christof Putzel, "Turning Rio's Favelas into a Tourist Destination," *Aljazeera America,* June 10, 2014, accessed February 28, 2015, http://america.aljazeera.com/watch/shows/america-tonight/articles/2014/6/10/turning-rio-s-favelasintoatouristattraction.html.

20. Marco De Nadai, Jacopo Satiano, Roberto Larcher, Nicu Sebe, Deniele Quercia, and Bruno Lepri, "The Death and Life of Great Italian Cities," *Proceedings of the 25th International Conference on the World Wide Web,* Montreal, Quebec, April 11–15, 2016 (Geneva, Switzerland: International WWW Steering Committee, 2016), 413–423; Vikas Mehta, "Lively Streets: Determining Environmental Characteristics to Support Social Behavior," *Journal of Planning Education and Research* 27, no. 2 (2007): 165–187; William H. Whyte, *The Social Life of Small Urban Spaces* (Washington, DC: Conservation Foundation, 1980).

21. Myrna M. Weissman and Eugene S. Paykel, "Moving and Depression in Women," in *Loneliness: The Experience of Emotional and Social Isolation,* ed. Robert S. Weiss (Cambridge, MA: MIT Press, 1973), 160.

22. Gans, *Levittowners,* 146.

23. See James Howard Kunstler, *The Geography of Nowhere* (New York: Simon and Schuster, 1993); John F. Freie, *Counterfeit Community* (Lanham, MD: Rowman and Littlefield, 1998).

24. Robert Gifford, "The Consequences of Living in High-Rise Buildings," *Architectural Science Review* 50, no. 1 (2007).

25. William L. Yancey, "Architecture, Interaction, and Social Control," *Environment, and Behavior* 3, no. 1 (1971): 3–21.

26. Margarethe Kusenbach, "A Hierarchy of Urban Communities: Observations on the Nested Character of Place," *City and Community* 7, no. 3 (2008): 225–249.

27. Hollie Lund, "Testing the Claims of New Urbanism," *Journal of the American Planning Association* 69, no. 4 (2003): 414–429; Lisa Wood, Lawrence D. Frank, and Billie Giles-Corti, "Sense of Community and Its Relationship with Neighborhood Design," *Social Science and Medicine* 70, no. 9 (2010): 1381–1390; Jacinta Francis et al., "Creating Sense of Community," *Journal of Environmental Psychology* 32, no. 4 (2012): 401–409.

28. See Jill Grant, *Planning the Good Community* (New York: Routledge, 2006); Emily Talen, "Sense of Community and Neighborhood Form," *Urban Studies* 36, no. 8 (1999): 1361–1379.

29. Oldenburg, *Great Good Place*, 215.

30. Grant, *Planning the Good Community*.

31. Jacobs, *Death and Life of Great American Cities*.

32. Gans, *Urban Villagers*; Oldenburg, *Great Good Place*.

33. Jacobs, *Death and Life of Great American Cities*, 74.

34. Gans, *Levittowners*, 145–146; cf. Robert D. Putnam, *Bowling Alone: The Collapse and Revival of American Community* (New York: Simon and Schuster, 2000).

35. Oldenburg, *Great Good Place*.

36. See Paul Hickman, "'Third Places' and Social Interaction in Deprived Neighborhoods in Great Britain," *Journal of Housing and the Built Environment* 28, no. 2 (2013): 221–236.

37. Robert G. Hollands, *Friday Night, Saturday Night: Youth Cultural Identification in the Post-Industrial City* (Newcastle-upon-Tyne, UK: University of Newcastle, 1995), 18.

38. Oldenburg, *Great Good Place*, 295–296.

39. Mario L. Small, *Villa Victoria* (Chicago: University of Chicago Press, 2005); also see Jill Grant, "Two Sides of the Same Coin? New Urbanism and Gated Communities," *Housing Policy Debate* 18, no. 3 (2007): 481–501; Gillette, *Civitas by Design*.

40. Gans, *Urban Villagers*.

41. See Roland L. Warren, *The Community in America* (Chicago, IL: Rand McNally, 1978).

42. See Evan McKenzie, *Beyond Privatopia: Rethinking Residential Private Government* (Washington, DC: Urban Institute Press, 2011).

43. Chris Morran, "HOA Bans Just About Everything Fun a Kid Might Do Outside," *Consumerist*, March 18, 2013, accessed February 28, 2015, https://consumerist .com/2013/03/18/hoa-bans-just-about-everything-fun-a-kid-might-do-outside.

44. Donald Shoup, *The High Cost of Free Parking* (Chicago, IL: APA Planners Press, 2005).

45. Lorene Hoyt, "The Business Improvement District Model," *Geography Compass* 1, no. 4 (2007): 946–958; Richard Briffault, "The Business Improvement District Comes of Age," *Drexel Law Review* 3, no. 19 (2010): 19–33.

46. Putnam, *Bowling Alone*, 213; Benjamin J. Newman, Joshua Johnson, and Patrick L. Lown, "The Daily Grind: Work, Commuting, and Their Impact on Political Participation," *American Politics Research* 42, no. 1 (2013): 141–170.

47. Eric J. Oliver, *Democracy in Suburbia* (Princeton, NJ: Princeton University Press, 2001).

48. See James DeFilippis, *Unmaking Goliath: Community Control in the Face of Global Capital* (New York: Routledge, 2004).

49. Suzanne Keller, *Community* (Princeton, NJ: Princeton University Press, 2005).

50. Baumgartner, *Moral Order of a Suburb*.

51. Quoted in Putnam, *Bowling Alone*, 210.

52. Setha Low, Gregory T. Donovan, and Jen Gieseking, "Shoestring Democracy," *Journal of Urban Affairs* 34, no. 3 (2012): 279–296.

53. Yancey, "Architecture, Interaction, and Social Control."

54. Charles Durrett and Kathryn McCamant, *Creating Cohousing* (Gabriola Island, BC: New Society Publishers, 2011); Dorit Fromm, "Seeding Community: Collaborative Housing as a Strategy for Social and Neighborhood Repair," *Built Environment* 38, no. 3 (2012): 364–394.

55. Jo Williams, "Designing Neighborhoods for Social Interaction: The Case of Cohousing," *Journal of Urban Design* 10, no. 2 (2005): 195–227.

56. Taylor Dotson, "Trial and Error Urbanism," *Journal of Urbanism* 9, no. 2 (2016): 148–165.

57. Baumgartner, *Moral Order of a Suburb*.

58. Gans, *Levittowners*, 142.

59. Erik Bichard, *The Coming of Age of the Green Community* (New York: Routledge, 2014).

60. See Michael Cheang, "Older Adults' Frequent Visits to a Fast-Food Restaurant: Nonobligatory Interaction and the Significance of Play in a 'Third Place,'" *Journal of Aging Studies* 16, no. 3 (2002): 303–321.

61. Wendell Berry, *What Matters? Economics for a Renewed Commonwealth* (Berkeley, CA: Counterpoint, 2010), 145.

62. Baumgartner, *Moral Order of a Suburb*; Regina Kenen, "Soapsuds, Spaces, and Sociability," *Journal of Contemporary Ethnography* 11, no. 2 (1982): 163–183.

Chapter 5

1. Richard E. Sclove, *Democracy and Technology* (New York: Guilford Books, 1995).

2. Adam Briggle and Carl Mitcham, "Embedding and Networking: Conceptualizing Experience in a Technosociety," *Technology in Society* 31, no. 4 (2009), 275, emphasis added.

3. See Kevin Lynch, *Good City Form* (Cambridge, MA: MIT Press, 1984).

4. Briggle and Mitcham, "Embedding and Networking," 380.

5. Taylor Dotson, "Technology, Choice, and the Good Life," *Technology in Society* 34, no. 4 (2012): 326–336.

6. David E. Nye, *America as Second Creation* (Cambridge, MA: MIT Press, 2004).

7. Stuart McLean (host), "Dave and the Elevator," *Vinyl Café*, CBC Radio, November 15, 2008, accessed February 25, 2011, https://itunes.apple.com/us/podcast/vinyl -cafe-stories-from-cbc/id263177347?mt=2.

8. Roberta M. Feldman and Susan Stall, "Resident Activism in Public Housing," in *Coming of Age*, ed. Robert I. Selby et al. (Urbana-Champaign, IL: Environmental Design Research Association, 1990), 111–119.

9. Albert Borgmann, *Technology and the Character of Contemporary Life* (Chicago: University of Chicago Press, 1984).

10. Christie Allen, "German Village Achieves Energy Independence … and Then Some," *BioCycle* 52, no. 8 (2011): 37–42.

11. Sara C. Bronin, "Curbing Energy Sprawl with Microgrids," *Connecticut Law Review* 43, no. 2 (2010): 547–584; Hannah J. Wiseman and Sara C. Bronin, "Community-Scale Renewable Energy," *San Diego Journal of Climate and Law* 14, no. 1 (2013): 1–29.

12. See Elinor Ostrom, *Governing the Commons: The Evolution of Institutions for Collective Action* (Cambridge, MA: Cambridge University Press, 1990); Donald Shoup, *The High Cost of Free Parking* (Chicago, IL: APA Planners Press, 2005).

13. See David Hess, *Localist Movements in a Global Economy* (Cambridge, MA: MIT Press, 2009); Gayle Christiansen, "Strengthening Small Business," in *Transforming Cities and Minds through the Scholarship of Engagement*, ed. Lorlene Hoyt (Nashville, TN: Vanderbilt University Press, 2013), 29–57.

14. Quoted in Barry Lynn, *Cornered: The New Monopoly Capitalism and the Economics of Destruction* (Hoboken, NJ: John Wiley and Sons, 2010), 115.

15. Troy Blanchard and Todd L. Matthews, "The Configuration of Local Economic Power and Civic Participation in the Global Economy," *Social Forces* 84, no. 4 (2006): 2241–2257; David S. Brown, "Discounting Democracy: Wal-Mart, Social Capital, Civic Engagement, and Voter Turnout in the United States" (working paper, Social Science Research Network, September 9, 2009), accessed September 1, 2014, http://ssrn.com/abstract=1398946; David S. Brown, Duncan Lawrence, and Anand Sokhey, "The Political Consequences of Big-Box Retail" (working paper, Annual Meeting of the Western Political Science Association, San Antonio, Texas, April 21–23, 2011), accessed September 1, 2014, https://duncanflawrence.files.wordpress.com/2012/09/wpsa-bigbox_efficacy1a-v2.pdf.

16. Matt Cunningham and Dan Houston, "The Civic Economics of Retail," *Civic Economics*, 2012, accessed September 1, 2014, http://nebula.wsimg.com/eb1a35cadd85dd440dcba5cb1eba005e?AccessKeyId=8E410A17553441C49302&disposition=0&alloworigin=1.

17. Robert D. Putnam, *Bowling Alone: The Collapse and Revival of American Community* (New York: Simon and Schuster, 2000).

18. Tim O'Brien, "Honest Weight Food Co-Op Discourages Union Drive," *Albany Times Union*, March 17, 2014, accessed June 12, 2014, http://www.timesunion.com/business/article/Honest-Weight-Food-Co-Opdiscourages-union-drive-5325871.php.

19. Wilson Majee and Ann Hoyt, "Are Worker-Owned Cooperatives the Brewing Pots of Social Capital?" *Community Development* 41, no. 4 (2010): 417–430.

20. Nick Iuviene and Lily Song, "Leveraging Rooted Institutions: A Strategy for Cooperative Economic Development in Cleveland, Ohio," in *Transforming Cities and Minds through the Scholarship of Engagement*, ed. Lorlene Hoyt (Nashville, TN: Vanderbilt University Press, 2013), 58–82; Steve Friess, "Can the Co-op Save Us?" *takepart*, May 30, 2014, accessed February 4, 2015. http://www.takepart.com/feature/2014/05/30/co-op-businesses-in-the-us-evergreen-cooperatives.

21. Jonathan Hall and Alan Krueger, "An Analysis of the Labor Market for Uber's Driver-Partners in the United States" (working paper 587, Industrial Relations Section, Princeton University, 2015).

22. See Sarah Kessler, "Pixel and Dimed," *Fast Company*, March 18, 2014, accessed June 12, 2014, https://www.fastcompany.com/3027355/pixel-and-dimed-on-not-gettingby-in-the-gig-economy.

23. Robert Kirkman, "At Home in the Seamless Web: Agency, Obduracy, and the Ethics of Metropolitan Growth," *Science, Technology, and Human Values* 34, no. 2 (2009): 43.

24. Michael W. Mehaffy, Sergio Porta, and Ombretta Romice, "The 'Neighborhood Unit' on Trial," *Journal of Urbanism* 8, no. 2 (2015): 199–217.

25. Donald Appleyard and Mark Lintell, "The Environmental Quality of City Streets," *Journal of the American Institute of Planners* 38, no. 2 (1972): 84–101.

26. William H. Whyte, *The Social Life of Small Urban Spaces* (Washington, DC: Conservation Foundation, 1980).

27. Zeynep Tufekci and Matthew E. Brashears, "Are We All Equally at Home Online?" *Information, Communication, and Society* 17, no. 4 (2014): 486–502.

28. Felicia W. Song, *Virtual Communities* (New York: Peter Lang, 2009).

29. Darin Barney, "The Vanishing Table, or Community in a World that Is No World," *TOPIA: Canadian Journal of Cultural Studies* 11 (2004): 49–66.

30. Keith Hampton and Barry Wellman, "Neighboring in Netville," *City and Community* 2, no. 3 (2003): 277–311.

31. Panayotis Antoniadis et al., "Community Building over Neighborhood Wireless Mesh Networks," *IEEE Technology and Society* 27, no. 1 (2008): 48–56; Primavera de Filippi, "It's Time to Take Mesh Networks Seriously," *Wired*, January 2, 2014, accessed June 12, 2014, https://www.wired.com/2014/01/its-time-to-take-mesh-networks-seriously-and-not-just-for-the-reasons-you-think/.

32. Gwen Shaffer, "Banding Together for Bandwidth: An Analysis of Survey Results from Wireless Community Network Participants," *First Monday* 16, no. 5 (2012), accessed June 12, 2014, http://pear.accc.uic.edu/ojs/index.php/fm/article/view/3331/2956.

33. Ishmael Johnathan et al., "Deploying Rural Community Wireless Mesh Networks," *IEEE Internet Computing* 12, no. 4 (2008): 22–29.

34. Heather Havrilesky, "Why Are Americans so Fascinated with Extreme Fitness?" *New York Times Magazine*, October 14, 2014, accessed October 14, 2014, http://www.nytimes.com/2014/10/19/magazine/why-are-americans-so-fascinated-with-extreme-fitness.html.

35. Jason Kessler, "Why I Quit CrossFit," *Medium*, July 15, 2013, accessed June 12, 2014, https://medium.com/this-happened-to-me/why-i-quit-crossfit-f4882edd1e21.

36. One only needs a few minutes perusing the "We Missed You" page on http://november-project.com to discover the creepy mixture of pseudosincerity and moralizing appeals to the sanctity of truth telling among November Project "friends."

37. Mathias Crawford, "Procedural Communities: Recreation Centers in Post-War America" (paper presented at the annual meeting of the Society for the Social Study of Science, San Diego, CA, October 9–12, 2013).

38. Troy D. Glover, "The 'Community' Center and the Social Construction of Citizenship," *Leisure Sciences* 26, no. 1 (2004): 63–83.

39. Heather Mair, "Club Life: Third Place and Shared Leisure in Canada," *Leisure Sciences* 31, no. 5 (2009): 450–465.

40. Stephen Krcmar, "Bending the Rules to Offer Yoga with a Beer Chaser," *New York Times*, June 11, 2014, accessed June 12, 2014, http://www.nytimes.com/2014/06/12/fashion/bending-the-rules-to-offer-yoga-with-a-beer-chaser.html.

41. Jonathan J. Brower, "The Professionalization of Youth Sport: Social Psychological Impacts and Outcomes," *Annals of the American Academy of Political and Social Science* 445, no. 1 (1979): 39–46.

42. See David C. Ogden, "Overgrown Sandlots: The Diminishment of Pickup Ball in the Midwest," *Nine* 10, no. 2 (2002): 120–130.

43. "Time in School: How Does the U.S. Compare?" Center for Public Education, accessed December 11, 2014, http://www.centerforpubliceducation.org/Main-Menu/Organizing-a-school/Time-in-school-How-does-the-US-compare.

44. Samuel Bowles and Herbert Gintis, "Schooling in Capitalist America Revisited," *Sociology of Education* 75, no. 1 (2002): 1–18; Peter Gray, *Free to Learn* (New York: Basic Books, 2013).

45. Jean M. Twenge and Joshua D. Foster, "Birth Cohort Increases in Narcissistic Personality Traits among American College Students, 1982–2009," *Social Psychological and Personality Science* 1, no. 1 (2010): 99–106; Sara H. Konrath, Edward H. O'Brien, and Courtney Hsing, "Changes in Dispositional Empathy in American College Students over Time: A Meta-Analysis," *Personality and Social Psychology Review* 15, no. 2 (2011): 180–198.

46. Gray, *Free to Learn*.

47. Renée Spencer, "Understanding the Mentoring Process Between Children and Adults," *Youth and Society* 37 (2006): 287–315; Chao Liu and Peter LaFreniere, "The Effects of Age-Mixing on Peer Cooperation and Competition," *Human Ethology Bulletin* 29, no. 1 (2014): 4–17; Mary Gordon, *The Roots of Empathy* (Toronto: Thomas Allen, 2005).

48. Robert Bellah et al. *Habits of the Heart.* (Berkeley: University of California Press, 1985).

49. John F. Freie, *Counterfeit Community: The Exploitation of Our Longings for Connectedness* (Lanham, MD: Rowman and Littlefield, 1998).

50. Justin G. Witford, *Sacred Subdivisions: The Postsuburban Transformation of American Evangelicalism* (New York: New York University Press, 2012).

51. Amy Florian, "My Neighbor, the Evangelical Megachurch," *Liturgy* 19, no. 4 (2004): 26.

52. See Witford, *Sacred Subdivisions*; Patricia Leigh Brown, "Megachurches as Minitowns," *New York Times*, May 9, 2002, accessed February 12, 2015, http://www.nytimes.com/2002/05/09/garden/megachurches-as-minitowns.html.

53. Eric Klinenberg, *Heat Wave: A Social Autopsy of Disaster in Chicago* (Chicago: University of Chicago Press, 2002).

54. See Freie, *Counterfeit Community*; Richard Cimino, "Neighborhoods, Niches, and Networks: The Religious Ecology of Gentrification," *City and Community* 10, no. 2 (2011): 157–181.

55. Pew Research Center, "'Nones' on the Rise," October 9, 2012, accessed March 2, 2015, http://www.pewforum.org/2012/10/09/nones-on-the-rise.

56. Ibid.; Pew Research Center, "Millennials in Adulthood," March 7, 2014, accessed March 2, 2015, http://www.pewsocialtrends.org/2014/03/07/millennials-in-adulthood.

57. David T. Beito, *From Mutual Aid to the Welfare State* (Chapel Hill: University of North Carolina Press, 2000).

58. Suresh Kumar and Mathews Numpeli, "Neighborhood Network in Palliative Care," *Indian Journal of Palliative Care* 11, no. 1 (2005): 6–9; Economist Intelligence Unit, *The Quality of Death*, 2010 report, *Economist*, accessed March 2, 2015, http://graphics.eiu.com/upload/eb/qualityofdeath.pdf.

59. Juan-Luis Klein et al., "The Québec Model," in *The International Handbook on Social Innovation*, ed. Frank Moulaert et al. (Northhampton, MA: Edward Elgar, 2013), 371–383.

60. Putnam, *Bowling Alone*.

61. Bob Solomon, "The Fall (and Rise?) of Community Banking," *UC Irvine Law Review* 2, no. 3 (2012): 945–983.

62. Bellah et al., *Habits of the Heart*.

63. Paul Lichterman, *The Search for Political Community* (New York: Cambridge University Press, 1996).

64. David Imbroscio, "Beyond Mobility: The Limits of Liberal Urban Policy," *Journal of Urban Affairs* 34, no. 1 (2012): 1–20.

65. Michael Gurstein, "Toward a Conceptual Framework for a Community Informatics," in *Connecting Canadians: Investigations in Community Informatics*, ed. A. Clement et al. (Edmonton, AB: Athabasca University Press, 2012), 35–60.

66. Sam Biddle, "Uber Cuts Prices in New York (and Fares for Drivers)," *ValleyWag*, July 7, 2014, accessed August 4, 2014, http://valleywag.gawker.com/uber-cuts-prices -in-new-york-and-fares-for-drivers-1601142883; Sam Biddle, "If TaskRabbit is the Future of Employment, the Employed are Fucked," *ValleyWag*, July 23, 2014, accessed August 4, 2014, http://valleywag.gawker.com/if-taskrabbit-is-the-future-of-employment-the-employed-1609221541.

Chapter 6

1. Arthur M. Schlesinger, "Biography of a Nation of Joiners," *American Historical Review* 50, no. 1 (1944): 24.

2. Albert Borgmann, *Technology and the Character of Contemporary Life* (Chicago: University of Chicago Press, 1984).

3. Matthew B. Crawford, *Shopclass as Soulcraft* (New York: Penguin, 2009).

4. Barbara E. Harrington et al., "Keeping Warm and Staying Well: Findings from the Qualitative Arm of the Warm Homes Project," *Health and Social Care in the Community* 13, no. 3 (2005): 263; also see Lars Kjerulf Petersen, "Autonomy and Proximity in Household Heating Practices," *Journal of Environmental Policy and Planning* 10, no. 4 (2008): 423–438; Mikko Jalas and Jenny Rinkinen, "Stacking Wood and Staying Warm: Time, Temporality and Housework around Domestic Heating Systems," *Journal of Consumer Culture*, November 11, 2013, accessed July 11, 2014, doi:10.1177/1469540513509639.

5. Lawrence E. Williams and John A. Bargh, "Experiencing Physical Warmth Promotes Interpersonal Warmth," *Science* 322, no. 5901 (2008): 606–607; Hans IJzerman and Gün R. Semin, "The Thermometer of Social Relations: Mapping Social Proximity on Temperature," *Psychological Science* 20, no. 10 (2009): 1214–1220.

6. Meryl Basham, Steve Shaw, and Andy Barton, *Central Heating: Uncovering the Impact on Social Relationships and Household Management* (Plymouth, UK: Plymouth and South Devon Research and Development Support Unit, 2004), accessed July 11, 2014, http://energybc.ca/cache/globalconsumereconomy/www.carillionenergy .com/downloads/pdf/central_heating.pdf.

7. Harold Wilhite et al., "A Cross-Cultural Analysis of Household Energy Behavior in Japan and Norway," *Energy Policy* 24, no. 9 (1996): 798.

8. Basham, Shaw, and Barton, *Central Heating*.

9. Darby Saxbe, Anthony Graesch and Marie Alvik, "Television as a Social or Solo Activity," *Communication Research Reports* 28, no. 2 (2011): 180–189.

10. Jason Reid, "'My Room! Private! Keep Out! This Means You!'" *Journal of the History of Childhood and Youth* 5, no. 3 (2012): 419–443.

11. Janet A. Flammang, *The Taste for Civilization: Food, Politics, and Civil Society* (Urbana: University of Illinois Press, 2009), 270.

12. John L. Locke, *The De-Voicing of Society* (New York: Simon and Schuster, 1998).

13. Sherry Turkle, "The Flight from Conversation," *New York Times*, April 21, 2012, accessed July 11, 2014, http://www.nytimes.com/2012/04/22/opinion/sunday/the -flight-from-conversation.html.

14. Jared Diamond, *The World until Yesterday* (New York: Viking, 2012).

15. Flammang, *Taste for Civilization*.

16. Eric Klinenberg, *Going Solo: The Extraordinary Rise and Surprising Appeal of Living Alone* (New York: Penguin Press, 2012).

17. Paul Lichterman, *The Search for Political Community* (New York: Cambridge University Press, 1996).

18. Tracy L. M. Kennedy and Barry Wellman, "The Networked Household," *Information, Communication, and Society* 10, no. 5 (2007): 645–670.

19. Eyal Ben-Ari, "'It's Bedtime' in the World's Urban Middle Classes: Children, Families and Sleep," in *Worlds of Sleep*, ed. L. Brunt and B. Steger (Berlin: Frank and Timme, 2008), 136–164; Diamond, *World until Yesterday*.

20. Ben-Ari, "It's Bedtime."

21. Robert Bellah et al., *Habits of the Heart* (Berkeley: University of California Press, 1985).

22. Reid, "My Room!"

23. James Mollison, *Where Children Sleep* (London: Chris Boot, 2010).

24. Jean Baudrillard, *The System of Objects*, trans. James Benedict (Paris: Editions Gallimard, 1968; New York: Verso, 2005).

25. Cf. Mihaly Csikszentmihalyi and Eugene Rochberg-Halton, *The Meaning of Things* (New York: Cambridge University Press, 1981).

26. Juliet B. Schor, *Born to Buy* (New York: Scribner, 2004).

27. *Fight Club*, directed by D. Fincher (Los Angeles, CA: 20th Century Fox, 1999), DVD.

28. Diamond, *World until Yesterday*.

29. Meret A. Keller and Wendy A. Goldberg, "Co-Sleeping: Help or Hindrance for Young Children's Independence," *Infant and Child Development* 13, no. 5 (2004): 369–388; James J. McKenna and Thomas McDade "Why Babies Should Never Sleep Alone: A Review of the Co-Sleeping Controversy in Relation to SIDS, Bedsharing and Breastfeeding," *Paediatric Respiratory Review* 6, no. 2 (2005): 134–152.

30. Paul Okami, Thomas Weisener, and Richard Olmstead, "Outcome Correlates of Parent-Child Bedsharing," *Developmental and Behavioral Pediatrics* 23, no. 4 (2002): 244–253.

31. Sanford Gaster, "Urban Children's Access to Their Neighborhood," *Environment and Behavior* 23, no. 1 (1991): 70–85; Lindsay Hasluck and Karen Malone, "Location, Leisure, and Lifestyle: Young People's Retreat to Home Environments." in *Through the Eyes of the Child*, ed. Constance L. Shehan (Stamford, CT: JAI Press, 1999), 177–196; Mayer Hillman, John Adams, and John Whitelegg, *One False Move … A Study of Children's Independent Mobility* (London: Policy Studies Institute, 1990); Karen Malone, "The Bubble-Wrap Generation," *Environmental Education Research* 13, no. 4 (2007): 513–527.

32. Diamond, *World until Yesterday*.

33. Gracy Olmstead, "Parenting in an Age of Bad Samaritans," *American Conservative*, July 17, 2014, accessed July 22, 2014, http://www.theamericanconservative.com/2014/07/17/parenting-in-an-age-of-bad-samaritans.

34. Malone, "Bubble-Wrap Generation."

35. Hanna Rosin, "The Overprotected Kid," *Atlantic*, March 19, 2014, accessed July 11, 2014, http://www.theatlantic.com/features/archive/2014/03/hey-parents-leave-those-kids-alone/358631.

36. Robert D. Putnam, *Bowling Alone: The Collapse and Revival of American Community* (New York: Simon and Schuster, 2000), 228.

37. Jaye L. Derrick, Shira Gabriel, and Kurt Hugenberg, "Social Surrogacy: How Favored Television Programs Provide the Experience of Belonging," *Journal of Experimental Social Psychology* 45, no. 2 (2009): 352–362.

38. Barbara Klinger, *Beyond the Multiplex: Cinema, New Technologies, and the Home* (Berkeley: University of California Press, 2006), 25.

39. Mike Snider, "Cocooning: It's Back and Thanks to Tech, It's Bigger," *USA Today*, February 18, 2013, accessed July 11, 2014, http://www.usatoday.com/story/tech/personal/2013/02/15/internet-tv-super-cocoons/1880473/; Mark Koba, "Keeping Fans in the Stands Is Getting Harder to Do," *CNBC*, July 16, 2013, accessed July 27, 2016, http://www.cnbc.com/id/100886843.

40. See Dong-Hee Shin, "Defining Sociability and Social Presence in Social TV," *Computers in Human Behavior* 29, no. 3 (2013): 939–947.

41. Bureau of Labor Statistics, "American Time Use Survey Summary—2015 Results," news release, June 24, 2016, accessed July 27, 2016, http://www.bls.gov/news .release/atus.nr0.htm; Simon Chapple and Maxime Ladaique, *Society at a Glance 2009* (Paris: OECD Social Policy Division, 2009), 34.

42. Georgjeanna Wilson-Doenges, "Push and Pull Forces Away from Front Porch Use," *Environment and Behavior* 33, no. 2 (2001): 264–278.

43. Eric J. Oliver, *Democracy in Suburbia* (Princeton, NJ: Princeton University Press, 2001).

44. Diana Mok, Barry Wellman, and Juan Carrasco, "Does Distance Matter in the Age of the Internet?" *Urban Studies* 47, no. 13 (2010): 2747–2783.

45. See Mohan J. Dutta-Bergman, "The Antecedents of Community-Oriented Internet Use," *Journal of Computer-Mediated Communication* 11, no. 1 (2006): 97–113; Michael J. Stern, Alison E. Adams, and Jeffrey Boase, "Rural Community Participation, Social Networks, and Broadband Use," *Agricultural and Resource Economics Review* 40, no. 2 (2011): 158–171.

46. Norman H. Nie and Kristen Backor, "The Development of the Internet in Everyday Life in America," in *Fortschritte der politischen Kommunikationsforschung Festschrift für Lutz*, ed. Birgit Krause, Benjamin Fretwurst, and Jens Vogelgesang (Wiesenbad: VS Verlag für Sozialwissenschaften, 2007), 131–151; Bertil Vilhelmson and Eva Thulin, "Virtual Mobility, Time Use and the Place of the Home," *Tijschrift voor Economishe en Sociale Geografie* 99, no. 5 (2008): 602–618; Ben Veenhof et al., *How Canadians' Use of the Internet Affects Social Life and Civic Participation,* Connectedness Series Report no. 16 (Ottawa: Statistics Canada, 2008), accessed April 1, 2015, http://publications.gc.ca/collections/collection_2008/statcan/56F0004M/ 56f0004m2008016-eng.pdf; Michael J. Stern, "How Locality, Frequency of Communication and Usage Affect Modes of Communication within Core Social Networks," *Information, Communication, and Society* 11, no. 5 (2008): 591–616; Irena Stephanikova, Norman H. Nie, and Xiaobin He, "Time on the Internet at Home, Loneliness, and Life Satisfaction," *Computers in Human Behavior* 26, no. 3 (2010): 329–338.

47. See Dohyun Ahn and Dong-Hee Shin, "Is the Social Use of Media for Seeking Connectedness or for Avoiding Social Isolation?" *Computers in Human Behavior* 29, no. 6 (2013): 2453–2462; Junghyun Kim, Robert LaRose, and Wei Peng, "Loneliness as the Cause and the Effect of Internet Use," *CyberPsychology and Behavior* 12, no. 4 (2009): 451–455.

48. Andrew Feenberg, *Questioning Technology* (New York: Routledge, 1999).

49. Loïc Wacquant, "A Janus-Faced Institution of Ethnoracial Closure," in *Ghetto,* ed. Ray Hutchinson and Bruce D. Haynes (Boulder, CO: Westview Press, 2012), 1–32.

50. See Keith N. Hampton, Oren Livio, and Lauren Sessions Goulet, "The Social Life of Wireless Urban Spaces," *Journal of Communication* 60, no. 4 (2010): 701–722; Rich Ling, *New Tech, New Ties* (Cambridge, MA: MIT Press, 2008).

51. Keith N. Hampton, "Community and Social Interaction in the Wireless City," *New Media and Society* 10, no. 6 (2008): 831–850.

52. Katrina Schwartz, "What Happens When Teens Try to Disconnect from Tech for Three Days," *MindShift*, March 6, 2015, accessed March 9, 2015, http://blogs.kqed .org/mindshift/2015/03/turned-off-how-teens-respond-to-a-no-tech-challenge.

53. Adrian Dater, "Team Bonding Suffers in Tech Age," *Sports Illustrated*, August 25, 2011, accessed October 14, 2014, http://www.si.com/more-sports/2011/08/25/ technology-teambonding.

54. Rod Chester, "Smartphone Apps Cloak and Split Let You Dodge Your Frenemies in Real Life," news.com.au, June 28, 2014, accessed July 11, 2014, http://www.news .com.au/technology/gadgets/smartphone-apps-cloak-and-splitlet-you-dodge-your -frenemies-in-real-life/story-fn6vihic-1226969315694; Megan Garber, "Chili's Has Installed More Than 45,000 Tablets in Its Restaurants," *Atlantic*, June 16, 2014, accessed July 11, 2014, http://www.theatlantic.com/technology/archive/2014/ 06/chilis-is-installing-tablet-ordering-at-all-its-restaurants/372836; Bruce Horovitz, "Panera Goes to High-Tech Ordering," *USA Today*, May 13, 2014, accessed July 11, 2014, http://www.usatoday.com/story/money/business/2014/05/13/panera-bread -fast-food-restaurants/9036545.

55. See Lee Rainie and Barry Wellman, *Networked: The New Social Operating System* (Cambridge, MA: MIT Press, 2012).

56. Melanie C. Green and Timothy C. Brock, "Antecedents and Civic Consequences of Choosing Real versus Ersatz Social Activities," *Media Psychology* 11, no. 4 (2008): 566–592.

57. Sherry Turkle, *Alone Together* (New York: Basic Books, 2011), 1.

58. Angela R. Stroud, "Good Guys and Bad Guys: Race, Class, Gender, and Concealed Handgun Licensing" (PhD diss., University of Texas at Austin, 2012).

59. Geoffery Canada, *Fist, Stick, Knife, Gun: A Personal History of Violence* (Boston, MA: Beacon Press, 1995); David Hemenway, Mary Vriniotis, and Matthew Miller, "Is an Armed Society a Polite Society?" *Accident Analysis and Prevention* 38, no. 4 (2006): 687–695.

60. Naomi Lakritz, "Kalamazoo Police Officer's Letter to Editor about Handguns Points to Cultural Divide," *Calgary Herald*, August 8, 2012, accessed July 11, 2014, http://www.calgaryherald.com/opinion/oped/Lakritz+Kalamazoo+police+officer+let ter+editor+about+handguns+points+cultural+divide/7054368/story.html.

61. Stroud, "Good Guys and Bad Guys."

62. Ibid., 204.

63. Jennifer D. Carlson, "I Don't Dial 911," *British Journal of Criminology* 52, no. 6 (2012): 1113–1132.

64. See Victor E. Kappeler and Larry K. Gaines, *Community Policing: A Contemporary Perspective*, 6th ed. (New York: Elsevier, 2011).

65. Fritz Umbach, *The Last Neighborhood Cops: The Rise and Fall of Community Policing in New York Public Housing* (New Brunswick, NJ: Rutgers University Press, 2011).

66. Jane Jacobs, *The Death and Life of Great American Cities* (New York: Random House, 1961).

67. Julie Beck, "Married to a Doll," *Atlantic*, September 6, 2013, accessed July 11, 2014, http://www.theatlantic.com/health/archive/2013/09/married-to-a-doll-why -one-man-advocates-synthetic-love/279361.

68. See Taylor Dotson, "Authentic Virtual Others? The Promise of Post-Modern Technologies," *AI and Society* 29, no. 1 (2014): 11–21.

69. David Levy, *Sex and Love with Robots* (New York: HarperCollins, 2007).

70. Niall Griffiths, "The Joys of a Repair Café: Mad Max with a KitKat and a Nice Cup of Tea," *Guardian*, September 17, 2014, accessed September 19, 2014, https:// www.theguardian.com/lifeandstyle/2014/sep/17/joys-repair-cafe-mad-max-kitkat -nice-cup-of-tea.

71. Crawford, *Shopclass as Soulcraft*.

72. E. F. Schuhmacher, *Small is Beautiful* (London: Blond and Briggs, 1973; New York: Harper Perennial, 2010).

73. Marianne De Laet and Anne Marie Mol, "The Zimbabwe Bush Pump," *Social Studies of Science* 30, no. 2 (2000): 225–263.

74. See Barry Schwartz, *The Paradox of Choice: Why Less Is More* (New York: Harper-Collins, 2004).

75. Quoted in Jason Wiener, "Ethical Architecture: UM Philosopher Albert Borgmann on the Good Life," *Missoula Independent*, February 15, 2007, accessed July 11, 2014, http://missoulanews.bigskypress.com/missoula/ethicalarchitecture/Content ?oid=1137832.

76. Shigehiro Oishi and Jason Kisling, "The Mutual Consitution of Residential Mobility and Individualism," in *Understanding Culture*, ed. Robert S. Wyer, Chi-yue Chiu, and Ying-yi Hong (New York: Psychology Press, 2009), 223–238; Shigehiro Oishi et al., "Residential Mobility Breeds Familiarity-Seeking," *Journal of Personality and Social Psychology* 102, no. 1 (2012): 149–162; Gundi Knies, "Neighborhood Social Ties: How Much Do Residential, Physical and Virtual Mobility Matter?" *British Journal of Sociology* 64, no. 3 (2013): 425–452.

Chapter 7

1. See Edward Younkins, "Technology, Progress, and Freedom," *Foundation for Economic Education*, January 1, 2000, accessed July 29, 2016. https://fee.org/articles/technology-progress-and-freedom; such Panglossian views of the market have been disputed by Geoffrey M. Hodgson, *Economics and Evolution* (Ann Arbor, MI: University of Michigan Press, 1996), among others.

2. See Anique Hommels, *Unbuilding Cities: Obduracy in Urban Socio-Technical Change.* (Cambridge, MA: MIT Press, 2005); Thomas P. Hughes, "Technological Momentum," in *Does Technology Drive History?*, ed. Merritt Roe Smith and Leo Marx (Cambridge, MA: MIT Press, 1994), 101–114, for a more complete introduction to the concepts of obduracy and technological momentum.

3. Edward J. Woodhouse, "Nanoscience, Green Chemistry, and the Privileged Position of Science," in *The New Political Sociology of Science*, ed. S. Frickel and K. Moore (Madison: University of Wisconsin Press, 2006), 148–181.

4. See Edward J. Woodhouse, "(Re)Constructing Technological Society by Taking Social Construction Even More Seriously," *Social Epistemology* 19, no. 2–3 (2005): 199–223.

5. See Sergio Sismondo, "Science and Technology Studies and an Engaged Program," in *Handbook of Science and Technology Studies*, 3rd ed., ed. Edward Hackett et al. (Cambridge, MA: MIT Press, 2007), 13–32.

6. For example, Robert D. Putnam, *Bowling Alone: The Collapse and Revival of American Community* (New York: Simon and Schuster, 2000).

7. The following works have inspired me in this regard: Charles E. Lindblom, "Who Needs What Social Research for Policymaking?" *Science Communication* 7, no. 4 (1986): 345–366; Bent Flyvbjerg, *Making Social Science Matter: Why Social Inquiry Fails and How It Can Succeed Again* (New York: Cambridge University Press, 2001).

8. Howard Davis, *The Culture of Building* (New York: Oxford University Press, 2006).

9. See Pamela Blais, *Perverse Cities: Hidden Subsidies, Wonky Policy, and Urban Sprawl* (Vancouver, Canada: UBC Press, 2010); Stefan Siedentop and Stefan Fina, "Who Sprawls the Most? Exploring the Patterns of Urban Growth across 26 European Countries," *Environment and Planning A* 44, no. 11 (2012): 2765–2784; Bianca Bosker, "Why Haven't Chinese Cities Learned from America's Mistakes?" *Guardian*, August 20, 2014, accessed November 6, 2014, https://www.theguardian.com/cities/2014/aug/20/why-havent-chinas-cities-learned-from-americas-mistakes.

10. See Charles E. Lindblom, "The Privileged Position of Business," in *The Big Business Reader on Corporate America*, ed. Mark Green (New York: Pilgrim Press, 1983): 193–203.

11. Blais, *Perverse Cities.*

12. Thad Williamson, David Imbroscio, and Gar Alperovitz, *Making a Place for Community* (New York: Routledge, 2002); Miriam Hortas-Rico and Albert Solé-Ollé, "Does Urban Sprawl Increase the Costs of Providing Local Public Services? Evidence from Spanish Municipalities," *Urban Studies* 47, no. 7 (2010): 1513–1540.

13. Christopher B. Leinberger, *The Option of Urbanism* (Washington DC: Island Press, 2008), 162.

14. See Arthur C. Nelson et al., *A Guide to Impact Fees and Housing Affordability* (Washington, DC: Island Press, 2008).

15. Blais, *Perverse Cities*; Jack Goodman, "Houses, Apartments, and the Incidence of Property Taxes," *Housing Policy Debate* 17, no. 1 (2006): 1–26.

16. John R. Logan and Harvey L. Molotch, *Urban Fortunes: The Political Economy of Place* (Berkeley University of California Press, 1987).

17. Charles Marohn, "The Growth Ponzi Scheme," *Strong Towns*, accessed February 25, 2015, http://www.strongtowns.org/the-growth-ponzi-scheme; in many municipalities, the police help make up the budget shortfall through excessive fines for insignificant infractions, often targeting poor minorities disproportionately.

18. H. Spencer Banzhaf and Nathan Lavery, "Can the Land Tax Help Curb Urban Sprawl? Evidence from Growth Patterns in Pennsylvania," *Journal of Urban Economics* 67, no. 2 (2010): 169–179; Dan Sullivan, "Why Pittsburgh Real Estate Never Crashes," in *Fleeing Vesuvius*, ed. Richard Douthwaite and Gillian Fallon (Gabriola Island, BC: New Society Publishers, 2011), 137–156.

19. Tony Dutzik, Benjamin David, and Phieas Baxandall, *Do Roads Pay for Themselves?* (Boston: U.S. PIRG, 2011); Joseph Henchman, "Gasoline Taxes and User Fees Pay for Half of State and Local Road Spending," *Tax Foundation*, January 3, 2014, accessed February 25, 2015, http://taxfoundation.org/article/gasoline-taxes-and-user-fees-pay-only-half-state-local-road-spending.

20. Juan Gomez and José Manuel Vassallo, "Comparative Analysis of Road Financing Approaches in Europe and the United States," *Journal of Infrastructure Systems* 20, no. 3 (2014), accessed April 1, 2015, http://ascelibrary.org/doi/pdf/10.1061/(ASCE)IS.1943-555X.0000193.

21. Donald Shoup, "The Trouble with Minimum Parking Requirements," *Transportation Research Part A: Policy and Practice* 33, no. 7–8 (1999): 549–574; Donald Shoup, *The High Cost of Free Parking* (Chicago, IL: APA Planners Press, 2005).

22. Tess Kalinkowski, "UP Express Fare to Include a Fee in Lieu of Airport Parking," *Toronto Star*, August 1, 2014, accessed August 6, 2014. http://www.thestar.com/news/gta/2014/08/01/up_express_fare_to_include_a_fee_in_lieu_of_airport_parking.html.

23. Shoup, *High Cost of Free Parking*.

24. Eric Toder et al., *Reforming the Mortgage Interest Deduction* (Washington, DC: Urban Institute, 2010).

25. Blais, *Perverse Cities*.

26. Rob Carrick, "Think Living in Suburbia's Cheaper? Think Again," *Globe and Mail*, October 7, 2013, accessed February 25, 2015, http://www.theglobeandmail. com/globe-investor/personalfinance/mortgages/thereal-cost-of-suburbia/ article14733912.

27. See Diane Francis, "The Chinese Are Coming, and They'd Like to Buy Your House," *Washington Post*, July 15, 2014, accessed November 5, 2014, https://www .washingtonpost.com/posteverything/wp/2014/07/15/the-chinese-are-coming-and -theyd-like-to-buy-your-house; James Surowiecki, "Real Estate Goes Global," *New Yorker*, May 26, 2014, accessed November 5, 2014, http://www.newyorker.com/ magazine/2014/05/26/real-estate-goes-global; Rachel Monroe, "More Guests, Empty Houses," *Slate*, February 13, 2014, accessed March 2, 2015, http://www.slate.com/ articles/business/moneybox/2014/02/airbnb_gentrification_how_the_sharing _economy_drives_up_housing_prices.html; Sam Roberts, "Homes Dark and Lifeless, Kept by Out-of-Towners," *New York Times*, July 6, 2011, accessed July 23, 2014, http://www.nytimes.com/2011/07/07/nyregion/more-apartments-are-empty-yet -rented-or-owned-census-finds.html; Frances Bula, "Vancouver's Vacancies Point to Investors, Not Residents," *Globe and Mail*, March 20, 2013, accessed July 23, 2014, http://www.theglobeandmail.com/news/british-columbia/vancouvers -vacanciespoint-to-investors-not-residents/article10044403.

28. Jessica Steinberg, "In Fight Against Jerusalem's 'Ghost Apartments,' Taxes May Not Do the Trick," *Times of Israel*, December 24, 2013, accessed July 23, 2014, http:// www.timesofisrael.com/doubled-property-tax-wont-solve-jerusalem-ghost-town -crisis.

29. David Linhart, "Eminent Domain Conversion of Vacant Luxury Condominiums into Low-Income Housing," *Boston University Public Interest Law Journal* 21, no. 1 (2011): 129–151.

30. European Environment Agency, *Urban Sprawl in Europe: The Ignored Challenge*, Report No. 10 (Copenhagen, Denmark: European Environment Agency/Office for Official Publication of the European Community, 2006).

31. Celia Barbero-Sierra, Maria-Jose Marques, and Manuel Ruíz-Pérez, "The Case of Urban Sprawl in Spain as an Active Agent Driving Force for Desertification," *Journal of Arid Environments* 90 (2013): 95–102.

32. See Joe R. Feagin and Robert E. Parker, *Building American Cities: The Urban Real Estate Game* (Washington, DC: BeardBooks, 1990).

33. John R. Logan and Harvey L. Molotch, *Urban Fortunes* (Berkeley: University of California Press, 1987); Gabor Zovanyi, *No-Growth Imperative* (New York: Routledge, 2013).

34. Colin Gordon, "Blight the Way: Urban Renewal, Economic Development, and the Elusive Definition of Blight," *Fordham Urban Law Journal* 31, no. 2 (2003): 305–337.

35. Greg LeRoy, "TIF, Greenfields, and Sprawl," *Planning and Environmental Law* 60, no. 2 (2008): 3–11.

36. Hommels, *Unbuilding Cities*.

37. Jill Grant, *Planning the Good Community* (New York: Routledge, 2006).

38. See Galina Tachieva, *Sprawl Repair Manual* (Washington, DC: Island Press, 2010); Ellen Dunham-Jones and June Williamson, *Retrofitting Suburbia* Hoboken, NJ: John Wiley and Sons, 2011).

39. See Leinberger, *Option of Urbanism*.

40. Jill Grant and Daniel E. Scott, "Complete Communities Versus the Canadian Dream," supplement, *Canadian Journal of Urban Research* 21, no. 1 (2012): 132–157.

41. Pierre Filion, "Reorienting Urban Development? Structural Obstructions to New Urban Forms," *International Journal of Urban and Regional Research* 34, no. 1 (2010): 1–19.

42. Seattle Urban Mobility Plan, "6 Case Studies in Urban Freeway Removal," 2008, accessed July 28, 2014, http://www.seattle.gov/transportation/docs/ump/06%20 seattle%20case%20studies%20in%20urban%20freeway%20removal.pdf.

43. Robert Cervero, Junhee Kang, and Kevin Shively, *From Elevated Freeways to Surface Boulevards* (Berkeley: University of California Transportation Center, 2007); Jason E. Billings, "The Impacts of Road Capacity Removal" (master's thesis, University of Connecticut, 2011).

44. Hommels, *Unbuilding Cities*, 38.

45. Canadian Mortgage and Housing Corporation, *Healthy High-Rise* (Ottawa, Canada: CMHC, 2001); Canadian Mortgage and Housing Corporation, "Using Building Form and Design," 2014, accessed February 25, 2015, https://www .cmhc-schl.gc.ca/en/inpr/afhoce/afhoce/afhostcast/afhoid/cohode/usbufode/; Peter Spurr, *Land and Urban Development* (Toronto: James Lorimer, 1976), 429; Oscar Newman, *Community of Interest* (Garden City, NY: Anchor Press/Doubleday, 1980), 117–120.

46. See Shoup, *High Cost of Free Parking*.

47. Donald Shoup, "In Lieu of Required Parking," *Journal of Planning Education and Research* 18 (1999): 307–320.

48. Shoup, *High Cost of Free Parking*.

49. Andrea Broaddus, "Tale of Two Ecosuburbs in Freiburg, Germany," *Transportation Research Record* 2187 (2010): 114–122.

50. See Emily Talen, *City Rules: How Regulations Affect Urban Form* (Washington, DC: Island Press, 2012); Emily Talen, "Zoning For and Against Sprawl: The Case for Form-Based Codes," *Journal of Urban Design* 18, no. 2 (2013): 175–200.

51. Town of Framingham, Massachusetts, *Zoning By-Law* (Framingham, MA: Framingham Planning Board, 2016), accessed December 13, 2016, http://www .framinghamma.gov/DocumentCenter/Home/View/24878.

52. See Chad D. Emerson, "Making Main Street Legal Again," *Missouri Law Review* 71, no. 3 (2006): 1–50.

53. Richard S. Geller, "The Legality of Form-Based Codes," *Journal of Land Use* 26, no. 1 (2010): 35–91.

54. Shoup, *High Cost of Free Parking*.

55. Talen, "Zoning For and Against Sprawl."

56. See Kenneth T. Jackson, *Crabgrass Frontier* (New York: Oxford University Press, 1985).

57. Grant, *Planning the Good Community*, 72.

58. Randall Arendt, "Planning Education: Striking a Better Balance," *Planetizen*, October 31, 2012, accessed July 25, 2014, http://www.planetizen.com/node/59072.

59. National Architecture Accreditation Board, *Conditions for Accreditation* (Washington, DC: NAAB, 2014), 16, accessed December 13, 2016, http://ced.berkeley.edu/ downloads/academic/accreditation/2014_NAAB_Conditions.pdf.

60. U.S. Census Bureau, "December 2013 Construction at $930.5 Billion Annual Rate," *U.S. Census Bureau News*, February 3, 2014, accessed July 22, 2014, http:// www.census.gov/const/C30/release.pdf.

61. Jennifer McCadney, "The Green Society? Leveraging the Government's Buying Powers to Create Markets for Recycled Products," *Public Contract Law Journal* 29, no. 1 (1999): 135–156.

62. Robert Kirkman, "At Home in the Seamless Web: Agency, Obduracy, and the Ethics of Metropolitan Growth," *Science, Technology, and Human Values* 34, no. 2 (2009): 234–258.

63. See Grant, *Planning the Good Community*.

64. Feagin and Parker, *Building American Cities*, 219.

65. See James DeFilippis, *Unmaking Goliath: Community Control in the Face of Global Capital* (New York: Routledge, 2004).

66. Ibid.; Tour of DSNI.

67. Peter Ache and Micha Fedrowitz, "The Development of Co-Housing Initiatives in Germany," *Built Environment* 38, no. 3 (2012): 395–412.

68. Carolin Funck, Tsutomu Kawada, and Yoshimichi Yui, "Citizen Participation and Urban Development in Japan and Germany," in *Urban Spaces in Japan: Cultural and Social Perspectives*, ed. C. Brumann and E. Schulz (New York: Routledge, 2012), 106–124.

69. Eirini Kasioumi, "Sustainable Urbanism," *Berkeley Planning Journal* 24, no. 1 (2011): 108.

70. "Alcohol Consumption Factsheet," Institute of Alcohol Studies, accessed July 28, 2014, http://www.ias.org.uk/uploads/pdf/Factsheets/Alcohol%20consumption%20factsheet%20April%202014.pdf.

71. Laura Donnelly, "Pubs Demand Minimum Alcohol Price," *Telegraph*, April 11, 2013, accessed July 28, 2014, http://www.telegraph.co.uk/news/health/news/9987635/Pubs-demand-minimum-alcohol-price.html.

72. Rick Muir, *Pubs and Places: The Social Value of Community Pubs* (London: Institute for Public Policy Research, 2012).

73. Lydia Lee, "A Community Approach to Clearing Snow from Sidewalks," *CityLab*, February 6, 2015, accessed February 23, 2015, http://www.citylab.com/cityfixer/2015/02/a-community-approach-to-clearing-snow-from-sidewalks/385173.

74. Arthur C. Nelson, "Leadership in a New Era," *Journal of the American Planning Association* 72, no. 4 (2006): 393–409.

75. Patrick C. Doherty and Christopher B. Leinberger, "The Next Real Estate Boom," *Washington Monthly*, November/December 2010, accessed September 26, 2014, http://washingtonmonthly.com/features/2010/1011.doherty-leinberger.html.

76. CBC News, "Throw-Away Buildings: Toronto's Glass Condos," November 14, 2011, accessed August 5, 2014, http://www.cbc.ca/news/canada/toronto/throw-away-buildings-toronto-s-glasscondos-1.1073319.

77. See Filion, "Reorienting Urban Development?"

Chapter 8

1. M. Granger Morgan and Hisham Zerriffi, "The Regulatory Environment for Small Independent Micro-Grid Companies," *Electricity Journal* 15, no. 9 (2002): 52–57;

Douglas E. King, "Electric Power Micro-Grids: Opportunities and Challenges for an Emerging Distributed Energy Architecture" (PhD diss., Carnegie Mellon University, 2006).

2. Sopitsuda J. Tongsopit and Brent M. Haddad, "Decentralised and Centralised Energy," *International Journal of Global Energy Issues* 27, no. 3 (2007): 326.

3. Hannah J. Wiseman and Sara C. Bronin, "Community-Scale Renewable Energy," *San Diego Journal of Climate and Law* 14, no. 1 (2013): 1–29.

4. Genevieve Rose Sherman, "Sharing Local Energy Infrastructure" (master's thesis, Massachusetts Institute of Technology, 2012).

5. Zuyi Li and Jiachun Guo, "Wisdom about Age," *IEEE Power and Energy Magazine* 4, no. 3 (2006): 44–51.

6. Sanya Carley and Richard N. Andrews, "Creating a Sustainable U.S. Electricity Sector: The Question of Scale," *Policy Sciences* 45 (2012): 97–121.

7. Jeff St. John, "New York Plans $40M in Prizes for Storm-Resilient Microgrids," *Green Tech Media*, January 9, 2014, accessed August 12, 2014, http://www .greentechmedia.com/articles/read/new-york-plans-40m-in-prizes-for-storm -resilient-microgrids.

8. See Wiseman and Bronin, "Community-Scale Renewable Energy."

9. Chris Marnay et al., "Policy Making for Microgrids," *IEEE Power and Energy Magazine* 6, no. 3 (2008): 66–77.

10. Joseph Loerentzen, "Decentralized Denmark," *Cogeneration and On-Site Power Production*, November-December 2005: 67–75; Danish Energy Agency, *Energy Statistics 2012* (Copenhagen: Danish Energy Agency, 2012).

11. Kiley Kroh, "Boulder, Colorado Faces Key Vote in Fight Against Carbon Pollution," *Think Progress*, November 5, 2013, accessed August 12, 2014, https:// thinkprogress.org/climate/2013/11/05/2889161/vote-boulder-climate-change; Mark Chediak, Christopher Martin, and Ken Wells, "Utilities Feeling Rooftop Solar Heat Start Fighting Back," *Bloomberg*, December 26, 2013, accessed August 12, 2014, http://www.bloomberg.com/news/2013-12-26/utilities-feeling-rooftop-solar-heat -start-fighting-back.html.

12. See Jane Jacobs, *The Death and Life of Great American Cities* (New York: Random House, 1961); Gayle Christiansen, "Strengthening Small Business," in *Transforming Cities and Minds through the Scholarship of Engagement*, ed. Lorlene Hoyt (Nashville, TN: Vanderbilt University Press, 2013), 29–57.

13. See Dan Ariely, *Predictably Irrational: The Hidden Forces that Shape our Decisions* (New York: Harper Collins, 2008); Edward F. Fischer, *Aspiration, Dignity, and the Anthropology of Wellbeing* (Stanford, CA: Stanford University Press, 2014).

14. David Fettig, "Thomas J. Holmes on Wal-Mart's Location Strategy," *Fedgazette*, March 1, 2006, accessed September 25, 2014, https://www.minneapolisfed.org/publications/fedgazette/thomas-j-holmes-on-walmarts-location-strategy; AAA, *Your Driving Costs: How Much Are You Really Paying to Drive?* (Heathrow, FL: AAA Association Communication, 2013), accessed September 25, 2014, https://exchange.aaa.com/wp-content/uploads/2013/04/Your-Driving-Costs-2013.pdf; Emily Jane Fox, "Wal-Mart: The $200 Billion Grocer," *CNN Money*, January 31, 2013, accessed September 25, 2014. http://money.cnn.com/2013/01/31/news/companies/walmart-grocery/index.html.

15. Margaret Gillerman, "Walmart Store Gets Bridgeton's Approval," *St. Louis PostDispatch*, July 8, 2010, accessed September 1, 2014, http://www.stltoday.com/news/local/metro/walmart-store-gets-bridgeton-s-approval/article_8d158273-5237-5e5d-b0d8-fc23c94dc3ff.html; Mary Sparacello, "Kenner City Council Approves $1.4 Million in Tax Incentives for Target," *Times-Picayune*, May 20, 2010, accessed September 1, 2014, http://www.nola.com/politics/index.ssf/2010/05/kenner_city_council_approves_1.html; Good Jobs First, "Subsidy Tracker 3.0," accessed September 1, 2014, http://www.goodjobsfirst.org/subsidy-tracker.

16. Americans for Tax Fairness, *Walmart on Tax Day* (Washington, DC: Americans for Tax Fairness, 2014), accessed September 2, 2014, http://www.americansfortaxfairness.org/files/Walmart-on-Tax-Day-Americans-for-Tax-Fairness-1.pdf; Sylvia A. Allegretto et al., *Fast Food, Poverty Wages* (Berkeley: University of California Berkeley Labor Center, 2013), accessed September 2, 2014, http://laborcenter.berkeley.edu/pdf/2013/fast_food_poverty_wages.pdf.

17. See David Hess, *Localist Movements in a Global Economy* (Cambridge, MA: MIT Press, 2009); Thad Williamson, David Imbroscio, and Gar Alperovitz, *Making a Place for Community* (New York: Routledge, 2002); Joshua Jansa and Virginia Gray, "The Politics and Economics of Corporate Subsidies in the 21st Century," working paper, University of North Carolina, Chapel Hill, 2014, accessed September 3, 2014, http://sppc2014.indiana.edu/Papers/JansaandGray.pdf.

18. Louise Story, "As Companies Seek Tax Deals, Governments Pay High Price," *New York Times*, December 1, 2012, accessed September 3, 2014, http://www.nytimes.com/2012/12/02/us/how-local-taxpayers-bankroll-corporations.html; Greg LeRoy, *The Great American Jobs Scam* (San Francisco: BerretKoehler, 2005).

19. William F. Fox and Matthew N. Murray, "Do Economic Effects Justify the Use of Fiscal Incentives?" *Southern Economic Journal* 71, no. 1 (2004): 78–92; Dafina Schwartz, Joseph Pelzman, and Michael Keren, "The Ineffectiveness of Location Incentive Programs," *Economic Development Quarterly* 22, no. 2 (2008): 167–179; Michael J. Hicks and Michael LaFaive, "The Influence of Targeted Economic Development Tax Incentives on County Economic Growth," *Economic Development Quarterly* 25, no. 2 (2011): 193–205; James T. Bennett, *They Play, You Pay: Why Taxpayers*

Build Ballparks, Stadiums, and Arenas for Billionaire Owners and Millionaire Players (New York: Springer, 2012).

20. Vicki Been, "Community Benefits Agreements: A New Local Government Tool or Another Variation on the Exactions Theme?" *University of Chicago Law Review* 77, no. 1 (2011): 5–35; Alejandro E. Camacho, "Community Benefits Agreements: A Symptom, Not the Antidote, of Bilateral Land Use Regulation," *Brooklyn Law Review* 78, no. 2 (2013): 355–389.

21. See Mark Taylor, "A Proposal to Prohibit Industrial Relocation Subsidies," *Texas Law Review* 72 (1993–1994): 669–713.

22. Samuel Nunn and Mark S. Rosentraub, "Metropolitan Fiscal Equalization: Distilling Lessons from Four U.S. Programs," *State and Local Government Review* 28, no. 2 (1996): 90–102.

23. Christiansen, "Strengthening Small Business."

24. See Victoria Williams, "Small Business Lending in the United States, 2012," *Small Business Research Summary* 414 (2013): 1–2; Ann Marie Wiersch and Scott Shane, "Why Small Business Lending Isn't What It Used to Be," *Economic Commentary* 10 (2013): 1–4; James A. Wilcox, *The Increasing Importance of Credit Unions in Small Business Lending* (Washington, DC: Small Business Administration, Office of Advocacy, 2011); Rob Witherell, "An Emerging Solidarity: Worker Cooperatives, Unions, and the New Union Cooperative Model in the United States." *International Journal of Labour Research* 5, no. 2 (2013): 251–268.

25. See Gilles L. Bourque, Margie Mendell, and Ralph Rouzier, "Solidarity Finance," in *Innovation and the Social Economy*, ed. Marie J. Bouchard (Toronto, Canada: University of Toronto Press, 2013), 180–205; Centre RELIESS, *Chaniter de l'economie sociale*, Brief no. 9 (Montreal: RELIESS, 2013), accessed September 4, 2014, http://base.socioeco.org/docs/chantieren1365012425.pdf.

26. Witherell, "An Emerging Solidarity."

27. Amy Cortese, "A Town Creates Its Own Department Store," *New York Times*, November 12, 2011, accessed November 19, 2014, http://www.nytimes.com/2011/11/13/business/a-town-in-new-york-creates-its-own-department-store.html.

28. See Nick Iuviene and Lily Song, "Leveraging Rooted Institutions: A Strategy for Cooperative Economic Development in Cleveland, Ohio," in *Transforming Cities and Minds through the Scholarship of Engagement*, ed. Lorlene Hoyt (Nashville, TN: Vanderbilt University Press, 2013), 58–82.

29. See Hess, *Alternative Pathways*, for more on import substitution.

30. See Bradford C. Snell, *American Ground Transport*, Report to the Committee of the Judiciary (Washington, DC: United States Government, 1974); Martha J. Bianco,

"The Decline of Transit: A Corporate Conspiracy or Failure of Public Policy? The Case of Portland, Oregon," *Journal of Policy History* 9, no. 4 (1997): 450–474.

31. William Vickrey, "Optimal Transit Subsidy Policy," *Transportation* 9 (1980): 389–409; Harry Clements, "A New Way to Predict Transit Demand," *Transit Economics* 23, no. 4 (1997): 49–52.

32. William Vickrey, "The City as a Firm," in *Economics of Public Services*, ed. by Martin Feldstein and Robert Inman (New York: Halsted, 1977), 334–343; Vickrey, "Optimal Transit Subsidy Policy."

33. Keith Wardrip, *Public Transit's Impact on Housing Costs: A Review of the Literature* (Washington, DC: Center for Housing Policy, 2011), accessed November 19, 2014, http://www.reconnectingamerica.org/assets/Uploads/TransitImpactonHsgCostsfinal -Aug1020111.pdf.

34. Jeffery J. Smith and Thomas A. Gihring, "Financing Transit Systems through Value Capture," *American Journal of Economics and Sociology* 65, no. 3 (2006): 751–786; Francesca Medda, "Land Value Capture Finance for Transport Accessibility: A Review," *Journal of Transport Geography* 25 (2012): 154–161; Zhirong Zhao et al., "Value Capture for Transportation Finance," *Procedia—Social and Behavioral Sciences* 48 (2012): 435–448.

35. Santa Monica Municipal Code, chapter 9, section 73, 2013, accessed April 1, 2015, http://www.qcode.us/codes/santamonica.

36. Donald Shoup and Richard Willson, "Employer-Paid Parking: The Problem and Proposed Solutions," *Transportation Quarterly* 46, no. 2 (1992): 169–192.

37. See Ethan Elkind, *Back in the Fast Lane: How to Speed Public Transit Planning and Construction in California*, Pritzker Briefs no. 6 (Emmett Institute on Climate Change and the Environment, Los Angeles, CA, 2014).

38. Will Doig, "Should It Take Decades to Build a Subway?" *Salon*, February 4, 2012, accessed September 8, 2014, http://www.salon.com/2012/02/04/should_it_take _decades_to_build_a_subway.

39. Robert Cervero, "Bus Rapid Transit (BRT): An Efficient and Competitive Mode of Public Transit" (working paper 2013-01, Institute of Urban and Regional Development, Berkeley, University of California, 2013).

40. William Yardley, "Seattle, After Decade of Debate, Approves Tunnel," *New York Times*, August 18, 2011, accessed December 18, 2014, http://www.nytimes.com/ 2011/08/19/us/19seattle.html.

41. Eric Jaffe, "How Utah Turned its Unpopular Transit System into a Hit," *CityLab*, October 21, 2013, accessed September 8, 2014, http://www.citylab.com/cityfixer/ 2013/10/how-utah-turned-its-unpopular-public-transit-system-hit/7298/.

42. Ibid.

43. Elkind, *Back in the Fast Lane.*

44. See Primavera de Filippi, "It's Time to Take Mesh Networks Seriously," *Wired*, January 2, 2014, accessed on June 12, 2014, https://www.wired.com/2014/01/its-time-to-take-mesh-networks-seriously-and-not-just-for-the-reasons-you-think.

45. Bart Braem et al., "A Case for Research with and on Community Networks," *ACM SIGCOMM Computer Communication Review* 43, no. 3 (2013): 68–73.

46. Gwen Shaffer, "Banding Together for Bandwidth," *First Monday* 16, no. 5 (2012), accessed June 12, 2014, http://pear.accc.uic.edu/ojs/index.php/fm/article/view/3331/2956.

47. Miquel Oliver, Johan Zuidweg, and Michail Batikas, "Wireless Commons against the Digital Divide," in *IEEE International Symposium on Technology and Society*, ed. Katina Michael (Wollongong, Australia: IEEE, 2010), 457–465.

48. Ishmael Johnathan et al., "Deploying Rural Community Wireless Mesh Networks," *IEEE Internet Computing* 12, no. 4 (2008): 22–29.

49. See Jan Krämer, Lukas Wiewiorra, and Christof Weinhardt, "Net Neutrality: A Progress Report," *Telecommunications Policy* 37, no. 9 (2013): 794–813.

50. Robert W. McChesney, *Digital Disconnect: How Capitalism is Turning the Internet Against Democracy* (New York: New Press, 2014); Douglas Rushkoff, "The Next Net," *Shareable*, January 3, 2011, accessed February 25, 2015, http://www.shareable.net/blog/the-next-net.

51. Moosung Lee and Tom Friedrich, "The 'Smaller' the School, the Better?" *Improving Schools* 10, no. 3 (2007): 261–282.

52. Linda McNeil, *The Contradictions of School Reform: Educational Costs of Standardized Testing* (New York: Routledge, 2000).

53. Peter Gray and David Chanoff, "Democratic Schooling: What Happens to Young People Who Have Charge of Their Own Education?" 94, no. 2 (1986): 182–213.

54. William Nack and Lester Munson, "Out of Control," *Sports Illustrated* 93, no. 4 (2000): 86–96.

55. Gill Valentine, *Public Space and the Culture of Childhood* (Burlington, VT: Ashgate, 2004).

56. Theda Skocpol, *Diminished Democracy: From Membership to Management in American Civic Life* (Norman: University of Oklahoma Press, 2003).

57. See Robert Bellah et al., *Habits of the Heart.* (Berkeley: University of California Press, 1985).

58. Stuart Murray, *Post Christendom: Church and Mission in a Strange New World* (Carlisle: Paternoster Press, 2004); Eddie Gibbs and Ryan K. Bolger, *Emerging Church: Creating Christian Community in Postmodern Cultures* (Grand Rapids, MI: Baker, 2005).

Chapter 9

1. See Anique Hommels, *Unbuilding Cities: Obduracy in Urban Socio-Technical Change* (Cambridge, MA: MIT Press, 2005); Robert Kirkman, "At Home in the Seamless Web: Agency, Obduracy, and the Ethics of Metropolitan Growth," *Science, Technology, and Human Values* 34, no. 2 (2009): 234–258.

2. See Meryl Basham, Steve Shaw, and Andy Barton, *Central Heating: Uncovering the Impact on Social Relationships and Household Management* (Plymouth, UK: Plymouth and South Devon Research and Development Support Unit, 2004), accessed July 11, 2014, http://energybc.ca/cache/globalconsumereconomy/www.carillionenergy .com/downloads/pdf/central_heating.pdf.

3. Elizabeth Shove, "Social, Architectural and Environmental Convergence," in *Environmental Diversity in Architecture*, ed. K. Steemers and M.A. Steane (New York: Spon Press, 2004), 19–30.

4. See Mike Russo and Dan Smith, *Apples to Twinkies 2013: Comparing Taxpayer Subsidies for Fresh Produce and Junk Food* (Washington, DC: U.S. PIRG, 2013), accessed February 25, 2015, http://uspirg.org/sites/pirg/files/reports/Apples_to_Twinkies _2013_USPIRG.pdf; The Physicians Committee for Responsible Medicine, "Farm Bill 2008: Who Benefits?" 2008, accessed October 1, 2014, http://www.pcrm.org/health/ health-topics/farmbill2008/farm-bill-2008-who-benefits.

5. Allison Aubrey and Dan Charles, "2 for 1: Subsidies Help Food Stamp Recipients Buy Fresh Food," *89.3 KPCC*, October 5, 2014, accessed October 24, 2014, http:// www.scpr.org/news/2014/10/05/47181/2-for-1-subsidies-help-food-stamp -recipients-buy-f/.

6. Mary Clare Jalonick, "Pizza Is a Vegetable? Congress Says Yes," *Today: Health*, November 15, 2011, accessed October 1, 2014, http://www.today.com/id/45306416/ ns/today-today_health/t/pizza-vegetablecongress-says-yes/#.VCwcVmddUrW; Ron Nixon, "New Rules for School Meals Aim at Reducing Obesity," *New York Times*, January 25, 2012, accessed October 1, 2014, http://www.nytimes.com/2012/01/26/ us/politics/new-school-lunch-rules-aimed-at-reducing-obesity.html.

7. J. Amy Dillard, "Sloppy Joe, Slop, Sloppy Joe: How USDA Commodities Dumping Ruined the National School Lunch Program," *Oregon Law Review* 87 (2008): 221–257.

8. Meredith Goad, "Chef's Touch Transforms School Lunches in Maine," *Portland Press Herald*, August 25, 2013, accessed October 1, 2014, http://www.pressherald

.com/2013/08/25/chefs-touch-transforms-school-lunches-in-maine_2013-08-25; Eleanor Beardsley, "Chef Proves School Lunches Can Be Healthy, Cheap," *NPR*, July 2, 2008, accessed October 1, 2014, http://www.npr.org/templates/story/story .php?storyId=91687769.

9. USDA: Food and Nutrition Services, "Farm to School," 2014, accessed October 1, 2014, http://www.fns.usda.gov/farmtoschool/farm-school.

10. James J. McKenna and Thomas McDade "Why Babies Should Never Sleep Alone: A Review of the Co-Sleeping Controversy in Relation to SIDS, Bedsharing and Breastfeeding," *Paediatric Respiratory Review* 6, no. 2 (2005): 134–152.

11. American Academy of Pediatrics, Task Force on Sudden Infant Death Syndrome, "The Chaning Concept of Sudden Infant Death Syndrome," *Pediatrics* 116, no. 5 (2005): 1245–1255.

12. McKenna and McDade, "Why Babies Should Never Sleep Alone."

13. Eyal Ben-Ari, "From Mothering to Othering: Organization, Culture, and Nap Time in a Japanese Day-Care Center," *Ethos* 24, no. 1 (1996): 136–164; Diana Adis Tahhan, "Depth and Space in Sleep: Intimacy, Touch and the Body in Japanese Co-Sleeping Rituals," *Body and Society* 14, no. 4 (2008): 37–56.

14. Melissa M. Burnham, "Co-Sleeping and Self-Soothing During Infancy," in *The Oxford Handbook of Infant, Child, and Adolescent Sleep and Behavior*, ed. Amy R. Wolfson and Hawley E. Montgomery (New York: Oxford University Press, 2013), 127–139.

15. Sara Latz, Abraham W. Wolf, and Betsy Lozoff, "Cosleeping in Context: Sleep Practices and Problems in Young Children in Japan and the United States," *JAMA Pediatrics* 153, no. 4 (1994): 339–346; Chisato Kawasaki et al., "The Cultural Organization of Infants' Sleep," *Children's Environments* 11, no. 2 (1999): 135–141.

16. Gill Valentine, *Public Space and the Culture of Childhood* (Burlington, VT: Ashgate, 2004), 18; Warwick Cairn, *How to Live Dangerously: Why We Should All Stop Worrying, and Start Living* (London: MacMillan, 2008).

17. David Millward, "First Mobile Phone Threatens Young Pedestrians," *Telegraph*, June 18, 2013, accessed November 8, 2014, http://www.telegraph.co.uk/motoring/ news/10125402/First-mobile-phone-threatens-young-pedestrians.html.

18. Valentine, *Public Space and the Culture of Childhood*.

19. Daniel Romer et al., "Cultivation Theory: Its History, Current Status, and Future Directions," in *The Handbook of Media and Mass Communication Theory*, ed. Robert S. Fortner and P. Mark Fackler (Malden, MA: John Wiley and Sons, 2014), 115–136.

20. Marianne B. Staempfli, "Reintroducing Adventure into Children's Outdoor Play Environments," *Environment and Behavior* 41, no. 2 (2009): 268–280; Hanna Rosin,

"The Overprotected Kid," *Atlantic*, March 19, 2014, accessed July 11, 2014, http://www.theatlantic.com/features/archive/2014/03/hey-parents-leave-those-kids-alone/358631.

21. John G. U. Adams, "Risk Homeostasis and the Purpose of Safety Regulation," *Ergonomics* 31, no. 4 (1988): 407–428.

22. See Joe L. Frost, "Playgrounds," in *Early Childhood Education: An International Encyclopedia*, vol. 3, ed. R.S. New and M. Cochran (Westport, CT: Praeger, 2007), 641–645.

23. Robert Kirkman, *The Ethics of Metropolitan Growth* (New York: Continuum, 2010).

24. Quoted in Melanie Thernstrom, "The Anti-Helicopter Parent's Plea," *New York Times Magazine*, October 19, 2016, accessed December 16, 2016, http://www.nytimes.com/2016/10/23/magazine/the-anti-helicopter-parents-plea-let-kids-play.html?_r=1.

25. See C. J. Pascoe, *Dude You're a Fag* (Berkeley: University of California Press, 2007); Sarah R. Edwards, Kathryn A. Bradshaw, and Verlin B. Hinsz, "Denying Rape but Endorsing Forceful Intercourse," *Violence and Gender* 1, no. 4 (2014): 188–193.

26. See Joseph R. Fogarty and Marcia Spielholz, "FCC Cable Jurisdiction: From Zero to Plenary in Twenty-Five Years," *Federal Communications Law Journal* 37 (1985): 113–129.

27. M. A. Izquierdo et al., "Air Conditioning in the Region of Madrid, Spain," *Energy* 36, no. 3 (2011): 1630–1639.

28. Yolande Strengers, "Air-Conditioning Australian Households: The Impact of Dynamic Peak Pricing," *Energy Policy* 38, no. 11 (2010): 7312–7322.

29. Norman Pressman, *Northern Cityscape: Linking Design to Climate* (Yellowknife, NT: Winter Cities Association, 1995), 76.

30. Jack Byers, "The Privatization of Downtown Public Spaces," *Journal of Planning Education and Research* 17, no. 3 (1998): 189–205; Stephen Graham and Simon Marvin, *Splintering Urbanism* (New York: Routledge, 2001).

31. Pressman, *Northern Cityscape*.

32. Laura Ryser and Greg Halseth, "Institutional Barriers to Incorporating Climate Responsive Design in Commercial Redevelopment," *Environment and Planning B: Planning and Design* 35, no. 1 (2008): 34–55.

33. Jason Koebler, "FCC Cracks Down on Cell Phone 'Jammers,'" *U.S. News and World Report*, October 17, 2012, accessed October 10, 2014, http://www.usnews.com/news/articles/2012/10/17/fcc-cracks-down-on-cell-phone-jammers; Bootie Crosgrove-Mather, "Cell Phone Jammers Keep Peace," *CBS News*, October 19, 2004,

accessed October 10, 2014, http://www.cbsnews.com/news/cell-phone-jammers
-keep-peace; E. Divya and R. Aswin, "Design of User Specific Intelligent Jammer," in
1st International Conference on Recent Advances in Information Technology, ed. Chiran-
jeev Kumar and Haider Banka (Calcutta, India: IEEE Communication Society, 2012),
312–316.

34. Alison Spiegel, "This Coffee Machine Forces You to Make Friends Before It'll
Dispense Your Drink," *Huffington Post,* April 17, 2014, accessed October 14, 2014,
http://www.huffingtonpost.com/2014/04/17/coffee-connector_n_5159909.html.

35. Fritz Umbach, *The Last Neighborhood Cops: The Rise and Fall of Community Policing in New York Public Housing* (New Brunswick, NJ: Rutgers University Press, 2011).

36. Allison T. Chappell, "The Philosophical Versus Actual Adoption of Community
Policing: A Case Study," *Criminal Justice Review* 34, no. 1 (2009): 5–28; Danijela
Spasic, Sladjana Djuric, and Zelimir Kesetovic, "Community Policing and Local
Self-Government," *Lex Localis* 11, no. 3 (2013): 293–309.

37. Chappell, "Philosophical Versus Actual"; Umbach, *Last Neighborhood Cops.*

38. Christopher D. O'Connor, "Empowered Communities or Self-Governing
Citizens?" in *Social Control: Informal, Legal and Medical,* ed. James Chriss (Bingley,
UK: Emerald, 2010), 129–148.

39. Umbach, *Last Neighborhood Cops.*

40. Gilliam Harper Ice, "Daily Life in a Nursing Home: Has It Changed in 25 Years?"
Journal of Aging Studies 16, no. 4 (2002): 345–359.

41. Robert D. Putnam, *Bowling Alone: The Collapse and Revival of American Community* (New York: Simon and Schuster, 2000); Christina R. Victor et al., "The Prevalence of, and Risk Factors for, Loneliness in Later Life," *Aging and Society* 25, no. 6
(2005): 357–375; Christina R. Victor and Keming Yang, "The Prevalence of Loneliness Among Adults: A Case Study of the United Kingdom," *Journal of Psychology* 146,
nos. 1–2 (2012): 85–104.

42. Andrea M. Petriwskyj and Jeni Warburton, "Motivations and Barriers to Volunteering by Seniors," *International Journal of Volunteer Administration* 24, no. 6 (2007):
3–25.

43. Francis G. Caro and Scott A. Bass, "Receptivity to Volunteering in Immediate
Postretirement Period," *Journal of Applied Gerontology* 16, no. 4 (1997): 427–441.

44. Social Security Administration, "Measures of Central Tendency for Wage Data,"
2012, accessed October 17, 2014, https://www.ssa.gov/oact/cola/central.html.

45. Maurizio Antoninetti, "The Difficult History of Ancillary Units," *Journal of Housing for the Elderly* 22, no. 4 (2008): 348–375; CBS Minnesota, "Mpls. Considers Allowing 'Granny Flats' in Residential Areas," June 13, 2014, accessed March 3, 2015,

http://minnesota.cbslocal.com/2014/06/13/mpls-considers-allowing-granny-flats -in-residential-areas/; Chris Poole, "New In-Law Suite Rules Boost Affordable Housing in San Francisco," *Shareable*, June 10, 2014, accessed March 3, 2015, http://www .shareable.net/blog/new-in-law-suite-rules-boost-affordable-housing-in-san -francisco.

46. Canadian Mortgage and Housing Corporation, "Residential Rehabilitation Assistance Program (RRAP)—Secondary/Garden Suite (On-Reserve)," 2014, accessed October 17, 2014, http://www.cmhc-schl.gc.ca/en/ab/hoprfias/hoprfias_010.cfm.

47. Amy Goyer, "Multigenerational Living: On the Rise," *AARP Blog: Take Care*, August 10, 2012, accessed October 17, 2014, http://blog.aarp.org/2012/08/10/ multigenerational-living-on-the-rise.

48. Shannon E. Jarrott and Kelly Bruno, "Shared Site Intergenerational Programs: A Case Study," *Journal of Applied Gerontology* 26, no. 3 (2007): 254.

49. Susan Pinker, *The Village Effect* (New York: Spiegel and Grau, 2014).

50. Jarrott and Bruno, "Shared Site Intergenerational Programs."

51. See Jared Diamond, *The World until Yesterday* (New York: Viking, 2012); Peter Gray, *Free to Learn* (New York: Basic Books, 2013).

52. Aral Balkan, *Free Is a Lie*, YouTube video, 32:24, posted by The Next Web, April 24, 2014, accessed February 24, 2015, https://www.youtube.com/watch?v =upu0gwGi4FE.

53. Breanna Weston, "Non-Traditional Lending Libraries in Oregon," *Oregon Library Association Quarterly* 19, no. 2 (2013): 11–16.

54. Gökçe Esenduran, Eda Kemahhoglu-Ziya, and Jayashankar Swaminathan, "Product Take-Back Legislation and Its Impact on Recycling and Remanufacturing Industries," in *Sustainable Supply Chains*, ed. T. Boone, V. Jayaraman, and R. Ganeshan (New York: Springer, 2012), 129–148.

55. Gökçe Esenduran, Atalay Atasu, and Luk N. Van Wassenhove, "How Does Extended Producer Responsibility Fare when E-Waste Has Value?" (working paper, Georgia Tech, Altanta, 2013), accessed April 1 2015, http://www.corporate -sustainability.org/conferences/sixth-annual-research-conference/Atasu.pdf.

56. Edward Woodhouse and Daniel Sarewitz, "Science Policies for Reducing Social Inequalities," *Science and Public Policy* 34, no. 3 (2007): 139–150.

57. Taylor Dotson, "Technology, Choice, and the Good Life," *Technology in Society* 34, no. 4 (2012): 326–336.

58. See Juliet Schor, *Plenitude: The New Economics of True Wealth* (New York: Penguin Press, 2010); Gabor Zovanyi, *No-Growth Imperative: Creating Sustainable Communities Under Ecological Limits to Growth* (New York: Routledge, 2013).

Chapter 10

1. See Edward Tenner, *Why Things Bite Back* (New York: Vintage Books, 1997).

2. Neil Postman and Charles Weingartner, *Teaching as a Subversive Activity* (New York: Dell, 1969), xiii.

3. Joseph G. Morone and Edward J. Woodhouse, *Averting Catastrophe: Strategies for Regulating Risky Technologies* (Berkeley: University of California Press, 1986); David Collingridge, *The Management of Scale* (New York: Routledge, 1992).

4. Edward J. Woodhouse, "Sophisticated Trial and Error in Decision Making About Risk," in *Technology and Politics*, ed. M.E. Kraft and N. Vig (Durham, NC: Duke University Press, 1988), 208–223; also see Robert D. Putnam, *Bowling Alone: The Collapse and Revival of American Community* (New York: Simon and Schuster, 2000).

5. Daniel J. Boorstin, *Republic of Technology: Reflections on Our Future Community* (New York: Harper and Row, 1978), 6–7.

6. See Evgeny Morozov, *The Net Delusion* (New York: Public Affairs, 2011).

7. Langdon Winner, "Artifacts/Ideas and Political Culture," in *Society, Ethics and Technology*, 4th ed., ed. Morton E. Winston and Ralph D. Edelbach (Boston: Wadsworth, 2012), 83–89.

8. See Morone and Woodhouse, *Averting Catastrophe*; Aaron Wildavsky, *Searching for Safety* (Piscataway, NJ: Transaction Publishers, 1988); Collingridge, *Management of Scale*; Edward J. Woodhouse, *The Future of Technological Civilization*, accessed August 5, 2014, https://www.academia.edu/2910120/The_Future_of_Technological _Civilization.

9. See Langdon Winner, *The Whale and the Reactor* (Chicago: University of Chicago Press, 1986); Steve Talbott, *The Future Does Not Compute* (Sebastopol, CA: O'Reilly and Associates, 1995); Sherry Turkle, *The Second Self* (Cambridge, MA: MIT Press, 1995); Albert Borgmann, *Holding on to Reality: The Nature of Information at the Turn of the Century* (Chicago: University of Chicago Press, 1999).

10. Lee Siegel, *Against the Machine: Being Human in the Age of the Electronic Mob* (New York: Spiegel and Grau, 2008), 8.

11. See Zeynep Tufekci and Matthew E. Brashears, "Are We All Equally at Home Online?" *Information, Communication, and Society* 17, no. 4 (2014): 486–502.

12. See Keith Hampton and Barry Wellman, "Neighboring in Netville," *City and Community* 2, no. 3 (2003), 277–311.

13. Eric A. Morris, "From Horse Power to Horsepower to Processing Power," *Freakonomics*, December 11, 2012, accessed November 3, 2014, http://freakonomics. com/2012/12/11/from-horse-power-to-horsepower-to-processing-power; Patrick Lin,

"The Ethics of Autonomous Cars," *Atlantic*, October 8, 2013, accessed November 3, 2014, http://www.theatlantic.com/technology/archive/2013/10/the-ethics-of -autonomous-cars/280360; Adam Thierer and Ryan Hagemann, "Removing Road-blocks to Intelligent Vehicles and Driverless Cars" (working paper (Arlington, VA: Mercatus Center, George Mason University, 2014).

14. Brian Wynne, "Seasick on the Third Wave? Subverting the Hegemony of Propositionalism," *Social Studies of Science* 33, no. 3 (2003): 401–417; Frank Fischer, *Democracy and Expertise: Reorienting Policy Inquiry* (New York: Oxford University Press, 2009).

15. Congressional Budget Office, *Alternative Approaches to Funding Highways*, Publication 4090 (Washington, DC: Congress of the United States, 2011), accessed February 25, 2015, http://thehill.com/images/stories/blogs/flooraction/Jan2011/cboreport.pdf.

16. Paul F. Hanley and Jon G. Kuhl, "National Evaluation of Mileage-Based Charges for Drivers," *Transportation Research Record* 2221 (2011): 10–18.

17. See Eric Jaffe, "The First Look at How Google's Self-Driving Car Handles City Streets," *CityLab*, April 28, 2014, accessed November 3, 2014, http://www.citylab .com/tech/2014/04/first-look-how-googles-self-driving-car-handles-city-streets/ 8977.

18. Putnam, *Bowling Alone*; Ding Ding et al., "Driving: A Road to Unhealthy Lifestyles and Poor Health Outcomes," *PLOS One* 9, no. 6 (2014), accessed April 1, 2015, http://journals.plos.org/plosone/article?id=10.1371/journal.pone.0094602; Margo Hilbrecht, Bryan Smale, and Steven E. Mock, "Highway to Health? Commute Time and Well-Being among Canadian Adults," *World Leisure Journal* 56, no. 2 (2014): 151–163.

19. Ronald Wright, *A Short History of Progress* (Toronto: House of Anansi Press, 2004).

20. See Mark Dowie, "Pinto Madness," *Mother Jones*, September/October 1977, accessed November 3, 2014, http://www.motherjones.com/politics/1977/09/ pintomadness; Michelle Maynard, "Toyota Cited $100 Million Savings After Limiting Recall," *New York Times*, February 21, 2010, accessed November 3, 2014, http:// www.nytimes.com/2010/02/22/business/22toyota.html.

21. Gail S. Kelley, *Construction Law* (Hoboken, NJ: John Wiley and Sons, 2013).

22. Michael Willats, "Death by Reckless Design: The Need for Stricter Criminal Statutes for Engineering-Related Homicides," *Catholic University Law Review* 58, no. 2 (2009): 567–598.

23. Thad Williamson, David Imbroscio, and Gar Alperovitz, *Making a Place for Community* (New York: Routledge, 2002), 312.

24. Jack C. Swearengen and Edward J. Woodhouse, "Cultural Risks of Technological Innovation: The Case of School Violence," *IEEE Technology and Society* 20, no. 1 (2001): 15–28.

25. Sherry Turkle, *Alone Together* (New York: Basic Books, 2011), 11.

26. See Taylor Dotson, "Authentic Virtual Others? The Promise of Post-Modern Technologies," *AI and Society* 29, no. 1 (2014): 11–21.

27. See Jill C. Manning, "The Impact of Internet Pornography on Marriage and the Family: A Review of the Research," *Sexual Addiction and Compulsivity* 13, no. 23 (2006): 131–165; Amanda M. Maddox et al., "Viewing Sexually-Explicit Materials Alone or Together: Associations with Relationship Quality," *Archives of Sexual Behavior* 40, no. 2 (2011): 441–448; Kirk Doran and Joseph Price, "Pornography and Marriage," *Journal of Family and Economic Issues* 35, no. 4 (2014): 489–498.

28. Turkle, *Alone Together*, 125.

29. David Levy, *Sex and Love with Robots: The Evolution of Human-Robot Relationships* (New York: HarperCollins, 2007).

30. Quoted in Lauren Orsini, "Jibo's Cynthia Breazeal: Why We Will Learn to Love Our Robots," *Readwrite*, August 11, 2014, accessed November 7, 2014, http://readwrite.com/2014/08/11/jibo-cynthia-breazeal-robots-social.

31. Levy, *Sex and Love with Robots*.

32. Turkle, *Alone Together*.

33. Alex Bereson et al., "Despite Warnings, Drug Giant Took Long Path to Vioxx Recall," *New York Times*, November 14, 2004, accessed November 10, 2014, http://www.nytimes.com/2004/11/14/business/14merck.html.

34. Zeynep Tufecki, "Do all these people who write about fake & wasteful social media ever ponder if this relates to their choices in friends & media usage?" *Twitter*, July 20, 2014, accessed February 25, 2015, https://twitter.com/zeynep/status/490848704639680512.

35. See Natasha Dow Schüll, *Addiction by Design: Machine Gambling in Las Vegas* (Princeton, NJ: Princeton University Press, 2012), a particularly edifying study of how gambling technology is purposefully designed to induce more addictive behavior in users.

36. Reynol Junco and Shelia R. Cotton, "Perceived Academic Effects of Instant Messaging Use," *Computers and Education* 56, no. 2 (2011): 370–378.

37. Ian Tucker, "Evgeny Morovoz: 'We Are Abandoning All the Checks and Balances,'" *Guardian*, March 9, 2013, accessed November 10, 2014, http://www.theguardian.com/technology/2013/mar/09/evgeny-morozov-technology-solutionism-interview.

38. Jathan Sadowski, "Office of Technology Assessment: History, Implementation, and Participatory Critique," *Technology in Society* 42 (2015): 9–20.

39. National Highway Traffic Safety Administration, "Preliminary Statement of Policy Concerning Automated Vehicles," accessed February 25, 2015, http://www .nhtsa.gov/staticfiles/rulemaking/pdf/Automated_Vehicles_Policy.pdf.

40. Amitai Etzioni, "The Capture Theory of Regulations—Revisited," *Society* 46, no. 4 (2009): 319–323.

41. Barry Estabrook, "How Gas Drilling Contaminates Your Food," *Salon*, May 18, 2011, accessed February 19, 2015, http://www.salon.com/2011/05/18/fracking _food_supply; Center for Science in the Public Interest, "FDA Fails to Protect Americans from Dangerous Drugs and Unsafe Foods," June 27, 2006, accessed June 28, 2016. https://cspinet.org/new/200606271.html.

42. Charles E. Lindblom and Edward J. Woodhouse, *The Policy-Making Process*, 3rd ed. (Englewood Cliffs, NJ: Prentice-Hall, 1993), 85.

43. Jameson M. Wetmore, "Amish Technology: Reinforcing Values and Building Community," *IEEE Technology and Society* 26, no. 2 (2007): 10–21.

44. Benjamin Barber, *If Mayors Ruled the World* (New Haven, CT: Yale University Press, 2013).

45. Richard E. Sclove, "Town Meetings on Technology," in *Science, Technology, and Democracy*, ed. Daniel Lee Kleinman (Albany: SUNY Press, 2000), 33–48.

46. Marc A. Olshan, "The National Amish Steering Committee," in *The Amish and the State*, ed. Donald B. Kraybill (Baltimore, MD: Johns Hopkins University Press, 2003), 67–86.

47. Mattahias Schwartz, "Pre-Occupied," *New Yorker*, November 28, 2011, accessed June 1, 2016, http://www.newyorker.com/magazine/2011/11/28/pre-occupied; Alex Mallis and Lily Henderson, "A Look Inside Occupy Wall Street in 'Right Here All Over,'" *Atlantic*, October 7, 2011, accessed June 1, 2016, http://www.theatlantic. com/video/index/246354/a-look-inside-occupy-wall-street-in-right-here-all-over; Josh Sanbum, "Square Roots: How Public Spaces Helped Mold the Arab Spring," *Time*, May 17, 2011, accessed June 1, 2016, http://content.time.com/time/world/ article/0,8599,2071404,00.html;

48. Yotam Maron, "What Really Caused the Implosion of the Occupy Movement —An Insider's View," *Alternet*, December 23, 2015, accessed June 1, 2016, http:// www.alternet.org/occupy-wall-street/what-really-caused-implosion-occupy -movement-insiders-view.

49. This distinction separates Woodhouse and Morone's take on the subject from the later work of David Collingridge; see Collingridge, *Management of Scale*; Morone

and Woodhouse, *The Demise of Nuclear Energy? Lessons for the Democratic Control of Technology* (New Haven, CT: University of Yale Press, 1989).

50. See Sophie Bond and Michelle Thompson-Fawcett, "Public Participation and New Urbanism," *Planning Theory and Practice* 8, no. 4 (2007): 449–472; Jill Grant, *Planning the Good Community* (New York: Routledge, 2006).

51. Greg LeRoy, "TIF, Greenfields, and Sprawl," *Planning and Environmental Law* 60, no. 2 (2008): 3–11.

52. Stan Alcorn, "Trying to Solve Albuquerque's Sprawl by Building a Development the Size of Manhattan," *Fast Company*, April 11, 2013, accessed November 11, 2014, https://www.fastcoexist.com/1681723/trying-to-solve-albuquerques-sprawl-by -building-a-development-the-size-of-manhattan.

53. Omar Mouallem, "From Public Contest to Quiet Dealings, City of Edmonton's Revised Blatchford Plan Raises Questions," *Metro News*, June 13, 2014, accessed November 11, 2014, http://metronews.ca/voices/footnotes/1065126/when-thecity -of-edmonton-played-script-doctor-on-the-blatchford-development-they-cut-more -than-a-few-lines.

54. Grant, *Planning the Good Community*.

55. Emily Talen, "Affordability in New Urbanist Development," *Journal of Urban Affairs* 32, no. 4 (2010): 489–510.

56. Thomas Schroepfer and Limin Hee, "Emerging Forms of Sustainable Urbanism," *Journal of Green Building* 3, no. 2 (2008): 65–76; Erik Bichard, *The Coming of Age of the Green Community* (New York: Routledge, 2014); Taylor Dotson, "Trial and Error Urbanism," *Journal of Urbanism* 9, no. 2 (2016): 148–165.

57. See Eirini Kasioumi, "Sustainable Urbanism," *Berkeley Planning Journal* 24, no. 1 (2011): 108; Carolin Funck, Tsutomu Kawada, and Yoshimichi Yui, "Citizen Participation and Urban Development in Japan and Germany," in *Urban Spaces in Japan: Cultural and Social Perspectives*, ed. C. Brumann and E. Schulz (New York: Routledge, 2012), 106–124.

58. Peter Ache and Micha Fedrowitz, "The Development of Co-Housing Initiatives in Germany," *Built Environment* 38, no. 3 (2012): 395–412.

59. Harrison Fraker, *The Hidden Potential of Sustainable Neighborhoods* (Washington DC: MIT Press, 2013), 99–100.

60. Cf. Grant, *Planning the Good Community*.

61. Crystal Stewart, Christopher Curro, Greg Dunn, and Kirsten Dunn, *Guide to Starting a Locally-Scaled, Local Foods Cooperative* (Burlington, VT: Northeast SARE, 2011), accessed November 19, 2014, http://www.nesare.org/content/download/ 70608/1002536/file/Guidetostartingalocally-scaledlocalfoodscooperative.pdf;

Michael DeMasi, "Closed Troy Co-op Has $1.9M in Debt," *Albany Business Review*, October 17, 2011, accessed November 19, 2014, http://www.bizjournals.com/ albany/news/2011/10/17/closed-troy-food-co-op-has-19m-in-debt.html.

62. Troy Food Coop, "Building Committee Update," August 23, 2008, accessed November 19, 2014, http://troyfoodcoop.blogspot.com/2008_08_01_archive.html; Chris Churchill, "A Reason the Troy Co-op Failed," *Albany Times Union*, November 16, 2011, accessed November 19, 2014, http://blog.timesunion.com/realestate/ a-reason-the-troy-food-co-op-failed/10540; Churchill, "On the Market: The Building that Was the Troy Co-op," *Albany Times Union*, December 7, 2011, accessed November 19, 2014, http://blog.timesunion.com/realestate/on-the-market-the-building -that-was-the-troy-co-op/10880.

63. Stewart et al., *Locally-Scaled, Local Foods Cooperative*; Christopher Curro and Kristen Stewart, "If We Can Make It There: A Food Coop's Secrets to Success in a Small City," workshop presented at New York State Neighborhood Revitalization Conference, Troy, NY, September 21, 2013.

64. Curro and Stewart, "If We Can Make It There."

65. See Cliff Ellis, "The New Urbanism: Critiques and Rebuttals," *Journal of Urban Design* 7, no. 3 (2002): 261–291.

66. See Merrill Paterson, "Mr. Jefferson's 'Sovereignty of the Living Generation,'" *VQR*, Summer 1976, accessed November 24, 2014, http://www.vqronline.org/essay/ mr-jefferson%E2%80%99s-%E2%80%9Csovereignty-living-generation%E2 %80%9D.

67. Edward J. Woodhouse and Jeff Howard, "Stealthy Killers and Governing Mentalities," in *Killer Commodities*, ed. M. Singer and H. Baer (Lanham, MD: Rowman and Littlefield, 2008), 46.

68. See Leo Marx, "Does Improved Technology Mean Progress?" *Technology Review* 90, no. 1 (1987), 33–41; Jathan Sadowski and Evan Selinger, "Creating a Taxonomic Tool for Technocracy and Applying It to Silicon Valley," *Technology in Society* 38 (2014): 161–168.

69. Adam Thierer and Ryan Hagemann, "Removing Roadblocks to Intelligent Vehicles and Driverless Cars" (working paper, Arlington, VA: Mercatus Center, George Mason University, 2014).

70. See Sadowski and Selinger, "Creating a Taxonomic Tool for Technocracy," 166, for more on technocratic governing mentalities.

71. See Taylor Dotson, "Technological Determinism and Permissionless Innovation as Technocratic Governing Mentalities," *Engaging Science, Technology, and Society* 1 (2015): 98–120, accessed June 24, 2016, http://estsjournal.org/article/view/8/20.

72. Langdon Winner, "Upon Opening the Black Box and Finding It Empty," *Science, Technology, and Human Values* 18, no. 3 (1993): 376.

73. Cf. Gwen Ottinger, *Refining Expertise: How Responsible Engineers Subvert Environmental Justice Challenges* (New York: NYU Press, 2013), a too rare example of STS research with clear lessons regarding how to realize saner technological societies.

74. Langdon Winner, "Do Artifacts Have Politics?" *Daedalus* 109, no. 1 (1980): 135.

Chapter 11

1. James Isaac and Irwin Altman, "Interpersonal Processes in Nineteenth Century Utopian Communities: Shakers and Oneida Perfectionists," *Utopian Studies* 9, no. 1 (1998): 26–49.

2. Janet R. White, "Designed for Perfection: Intersections between Architecture and Social Program at the Oneida Community," *Utopian Studies* 7, no. 2 (1996): 113–138.

3. See Robert Bellah et al., *Habits of the Heart.* (Berkeley: University of California Press, 1985).

4. David Imbroscio, "Beyond Mobility: The Limits of Liberal Urban Policy," *Journal of Urban Affairs* 34, no. 1 (2012): 1–20.

5. M. P. Baumgartner, *The Moral Order of a Suburb* (New York: Oxford University Press, 1988); Eric Klinenberg, *Going Solo: The Extraordinary Rise and Surprising Appeal of Living Alone* (New York: Penguin Press, 2012).

6. See Thomas P. Hughes, "Technological Momentum," in *Does Technology Drive History?*, ed. Merritt Roe Smith and Leo Marx (Cambridge, MA: MIT Press, 1994), 101–114; Richard Perkins, "Technological 'Lock-In.'" *Internet Encyclopedia of Ecological Economics.* (International Society for Ecological Economics, 2003), accessed December 8, 2014, http://isecoeco.org/pdf/techlkin.pdf.

7. Edward J. Woodhouse and Jason W. Patton, "Design by Society: Science and Technology Studies and Social Shaping of Design," *Design Issues* 20, no. 3 (2004): 6; also Langdon Winner, *The Whale and the Reactor* (Chicago: University of Chicago Press, 1986).

8. See chapter 2; also Lee Rainie and Barry Wellman, *Networked: The New Social Operating System* (Cambridge, MA: MIT Press, 2012).

9. Edward J. Woodhouse, "(Re)Constructing Technological Society by Taking Social Construction Even More Seriously," *Social Epistemology* 19, no. 2–3 (2005): 199–223.

10. For example, Robert D. Putnam, *Bowling Alone: The Collapse and Revival of American Community* (New York: Simon and Schuster, 2000).

11. Jill Grant, *Planning the Good Community* (New York: Routledge, 2006).

12. Herbert J. Gans, *The Levittowners: Ways of Life and Politics in a New Suburban Community* (New York: Pantheon Books, 1967); Baumgartner, *Moral Order of a Suburb.*

13. See Joseph G. Morone and Edward J. Woodhouse, *Averting Catastrophe: Strategies for Regulating Risky Technologies* (Berkeley: University of California Press, 1986); David Collingridge, *The Management of Scale* (New York: Routledge, 1992).

14. Ray Oldenburg, *The Great Good Place* (1989; repr., Cambridge, MA: Da Capo Press, 1999), 247; also Naomi Gerstel and Natalia Sarkisian, "Marriage: The Good, the Bad, and the Greedy," *Contexts* 5, no. 4 (2006): 16–21.

15. Juliet Schor, *Plenitude: The New Economics of True Wealth* (New York: Penguin, 2010).

16. Edward J. Woodhouse, "Nanoscience, Green Chemistry, and the Privileged Position of Science," in *The New Political Sociology of Science,* ed. S. Frickel and K. Moore (Madison: University of Wisconsin Press, 2006), 148–181.

17. Judy Wajcman, *Pressed for Time* (Chicago: University of Chicago Press, 2014).

18. Ralph Nader, "Thanksgiving for Social Scientists," *Huffington Post,* November 26, 2014, accessed December 8, 2014, http://www.huffingtonpost.com/ralphnader/thanksgiving-for-social-s_b_6225936.html.

19. Bent Flyvbjerg, *Making Social Science Matter: Why Social Inquiry Fails and How It Can Succeed Again* (New York: Cambridge University Press, 2001).

Index

Printed in the United States
by Baker & Taylor Publisher Services